TABLE OF CONTENTS

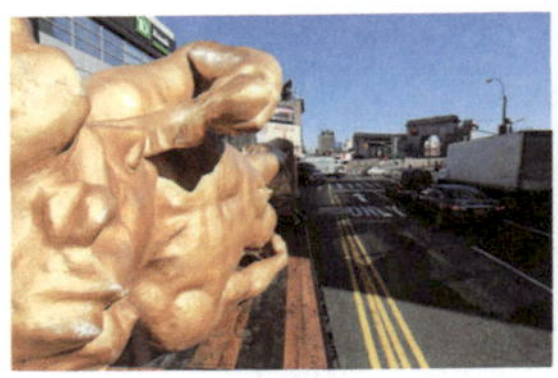

COLOPHON

The Manifesto Series is curated by Storefront for Art and Architecture
Eva Franch i Gilabert
Executive Director and Chief Curator
Carlos Mínguez Carrasco
Associate Curator
Kara Meyer
Director of External Relations
Zeynep Goksel
Development Associate
Piotr Chizinski
Gallery Manager and Project Coordinator

Editor
Serkan Özkaya

Copy Editor
Justin Fowler

Texts by
Ines Weizman, Fake Industries Architectural Agonism, Hrag Vartanian, Spyros Papapetros, Christopher Eamon, Alanna Heiss and Serkan Özkaya

Discussions with
Eva Franch i Gilabert, Ines Weizman, Cristina Goberna, Alanna Heiss, Spyros Papapetros, Hrag Vartanian, Christopher Eamon, Liza Béar, Pedro Gadanho, Urtzi Grau, Lynne Tillman, Michael Morris and Serkan Özkaya

Special Thanks to
21c Museum, Steve Wilson, Laura Lee Brown, Alice Gray Stites, Nico Jorcino, Sina Najafi, Brett Beyer, Baris Ozcetin, Gabriel Wrye, Bati Logistics and Gillian Steinhardt

Design by
Pentagram: Natasha Jen, Yen-Wei Liu

Printing
Regal Printing, Hong Kong

Lars Müller Publishers
Zürich, Switzerland
www.lars-mueller-publishers.com

ISBN 978-3-03778-345-0

Printed in Hong Kong

Every effort has been made to gain permission from copyright holders and / or photographers, where known, for the images reproduced in this book, and care has been taken to credit those images correctly. Any omissions are unintentional and we will be happy to include appropriate credits in any future editions if further information is brought to the publisher's attention.

Manifesto Series: Double was made possible by the New York State Council on the Arts and public funds from the New York City Department of Cultural Affairs in partnership with the City Council.

MANIFESTO FOR THE RIGHTS OF ARCHITECTURE

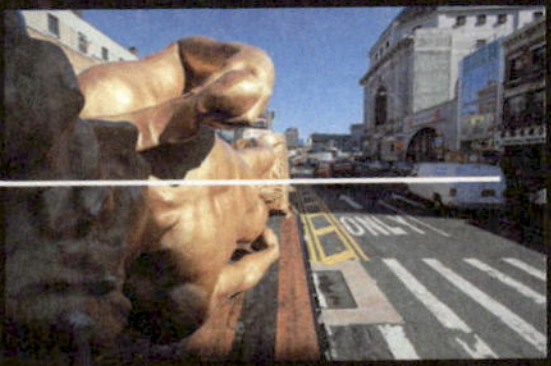

Ines Weizman

We often speak about human rights.

Some architects attempt to speak about the human-right violation by means of architecture. But what about the rights of architecture itself? Architectural rights? Can we inversely think of architectural-right-violation by humans?

Let's start somewhat "smaller" – with objects – with those things previously known as "Ding(s)." One of the most fundamental "rights of the object" is the right of reproduction.

In the object world, reproduction operates by the mechanisms of copying and variation. Regulating this form of reproduction is the right to copy – or copyright.

If copyright is to objects what human rights are for people, then the right of copying / reproduction will have to be rethought.

Copyright is currently understood as the consequence of determining the identity of the maker of an object, a thing, an idea, a structured assemblage, or a building.

Copyright is the right to copy, replicate, duplicate, and receive the financial benefits of this operation. It is the privilege of a maker.

But what if copyrights are not to be thought of as the rights of designers but as those of the objects / buildings themselves?

For such a shift to happen, we need to consider objects / buildings as something like subjects – the bearers of rights.

The law must treat objects / buildings in a way similar to how it treats children. Children have rights, but can't represent themselves. They are spoken for by their parents until they are eighteen.

Buildings, too, can be the bearers of duties, responsibilities, and liabilities. The most basic of all rights (and wrongs) – as we said – are the rights (and wrongs) of reproduction.

In this context, architecture today is at the height of the Victorian era – alternately prude, fearful, possessive, and submissive.

Today's form of architectural reproduction is highly regulated, censored, and belittling of its offspring.

Copyright in the way it is today conceived exists in the spectrum between what trends architects can follow – that which exerts influence or exists as a thing to be copied – and where the intellectual property of an individual begins (an area that should not be transgressed).

Copyright thus regulates the distance between "influence" and "rip off," "fake" and "theft." This distance, shifting and dynamic, subject to opening and closing, is a matter of a cultural, social, and political contract articulated and tested by law.

Every object is haunted by its potential or actual double (or doubles) which come into being as replicas, simulacras, reproductions, or fakes – a production series in which gradual differentiations and mutations may occur.

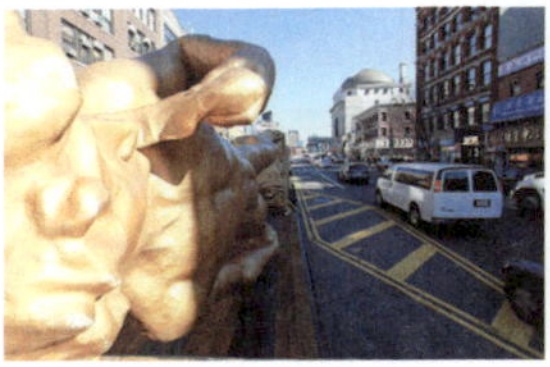

In the field of contemporary automata and semi-automata, architecture is produced in the digital and physical sphere precisely as systems of copy and reproduction in its components and in its entirety.

Rights and wrongs are those of the software, and they exist well in excess of the rights and wrongs of those people who participated in the diffuse and sometimes open source making of this software.

Architecture is, of course, a special kind of thing.

It is a thick layer of relations, references, technologies, services, and infrastructures that act and interact in the world.

It is, one could say, a quasi-subject. It has an agency although, perhaps, it does not possess human intentionality.

Architects often speak about the agency of the architect, or that of the stakeholders, but how might we think about the agency of architecture itself?

Just like the deity whose sculptural representation is a coming into being – and at a certain point exceeds the power of its maker – so, too, does architecture gain a certain power to act which is out of the hands of its maker.

This capacity is what Bruno Latour calls the "factish." Although fabricated, a thing can exercise its independent agency.

Isn't architecture thus the best example of the factish? Don't buildings have life and effect beyond their under-signed architect?

We often say that buildings behave and perform. Can they also mis-perform or misbehave? If this architecture – always in excess of its maker – has agency, it must also have liability and could be tried (and sometimes punished – even exiled or executed).

Contemporary architecture is a product of a growing force-field – financial, political, cultural, discursive, effective, and affective. It folds micro and macro forces into its organization and form. The architect's team translates some of these forces and is swayed by others, but this team is by no means the master of form as architecture schools would have us believe.

Foucault thought of authorship as a social construct – something that was developed in the 18th century – but hoped for a world in which it did not matter who the author of a text was.

Barthes, asserting that the text is merely a product of other texts, argues that authorship is to be replaced by inter-textuality.

We already heard of inter-architectuality! (But not the postmodern kind.)

So, where is the space in which the agency and authorship of architecture can perform?

Let's look at the institution of the court and the process of a trial as a place where architecture can perform.

The trials of inanimate objects have been long abolished. We know about ancient and medieval trials where sculptures, vehicles, building parts, and animals were tried and "executed." At present, these trials of things seem to return in the increasingly frequent debates around copyright laws.

In architecture, copies need a real or imaginary tabula rasa – "new worlds" – and indeed the new frontiers of this field of experimentation are in the sand dunes of the new territories in the Middle East (think Dubai) and China (think Inner Mongolia), but also elsewhere.

Remember the heated words exchanged between Koolhaas and Foster regarding the "remarkable similarity" – gridded, square-shaped, sun-powered, self-sustained eco-cities – that defined both Foster's Masdar in the outskirts of Abu Dhabi and Koolhaas's RAK Gateway in the UAE? Or Tschumi and Eisenman clashing over the grid of La Villette?

The paradox with practices such as Foster's and Rem's is that the more the design process becomes diffused, multidisciplinary, open ended, and open sourced, the greater the desire becomes to defend a single brand-name.

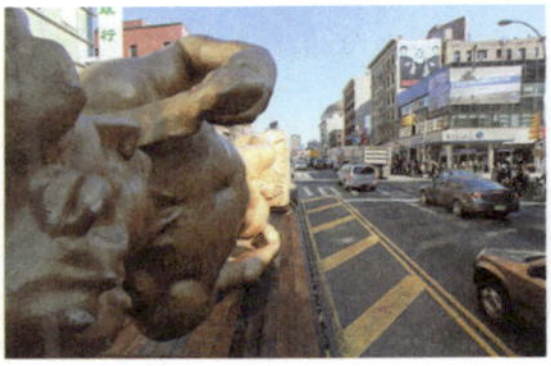

Architectural copyright cases rarely go to trial. Shame. Trials could become the most important arenas in which architecture is interrogated and debated and allowed to perform.

We could imagine, in the OMA / Foster case, buildings like the skyscrapers in Madelon Vriesendorp's illustrations: buildings after making love, getting out of bed, standing up, presenting the case, being called to testify, performing cross-examinations, rendering verdicts amongst themselves...?

Architectural representations – drawings, models, animations – are not presented, but presenting; not disputed as evidence in a courtroom, but disputing. The objects of architectural representation turn the courtroom into a critical arena of architectural debate – the arena that once was urbanism. Could the city turn into a courtroom?

In Bolivia, so a friend recently mentioned, the rights of

rivers – to be clean, for example – are now protected by law regardless of the effects they exercise on humans. Nature will be protected on its own terms.

What are the benefits of copyrights – of rights being redefined as those of buildings and not of makers?

Simply: the discussion of architecture could free itself from the discussion of the architect, and shift to the discussion of architecture itself.

Departing from the endless pseudo-psychoanalytic period we would enter one of super-objectivity.

The rights of the object must exist outside of the rights, reputation, and financial interests of the architect.

Sometimes architecture has to be protected from its architect.

Copying – and the associated morphological processes and distortions – is the way in which buildings / things reproduce themselves – and for buildings / things to reproduce themselves they do need humans.

Buildings need to use humans with their competitive and creative spirit, their ingenuity and laziness, their vanity and inclination to corruption – and this just as we use prostheses, like we use computers and we use cars, telephones, pace-makers, and walking sticks.

Objects / buildings in the process of reproducing themselves need humans like humans need a bottle of wine... Can we step down from our mastery and start understanding our role as being no longer the sole creators and prude regulators of architecture and finally start acting like we are this necessary half a bottle of wine?

Example

I have undertaken a project that tests and challenges the notion of whose rights are copyrights.

I have been invited alongside 99 others to build a villa in Ordos. The brief asked for a 10,000 square feet villa and a swimming-pool. I proposed to build Loos's 1927 unrealised project for a house for Josephine Baker. The year the project was announced was the 75th anniversary of the year of the death of Adolf Loos, 1933.

Copyrights were no longer protected in the sense of belonging to him or to a relative, but they still belonged to the building.

They were now the rights of the building itself.

These rights are not simply a charge for endless self-replication.

They have, instead, to be carefully considered apart from the interests of Loos or of any of his relatives.

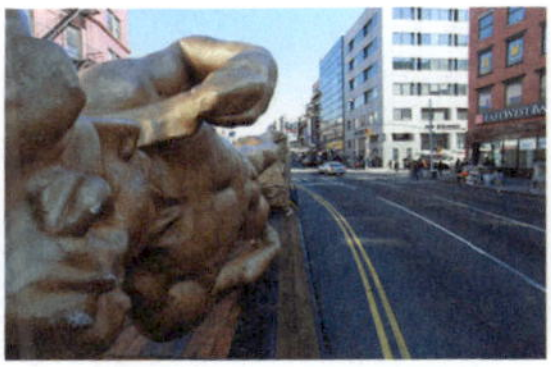

In the work of designing a copy without an original, it turned out that an architectural historian and expert in the work of Adolf Loos and a copyright lawyer would be the ones to help judge whether or not I have violated the rights of the building.

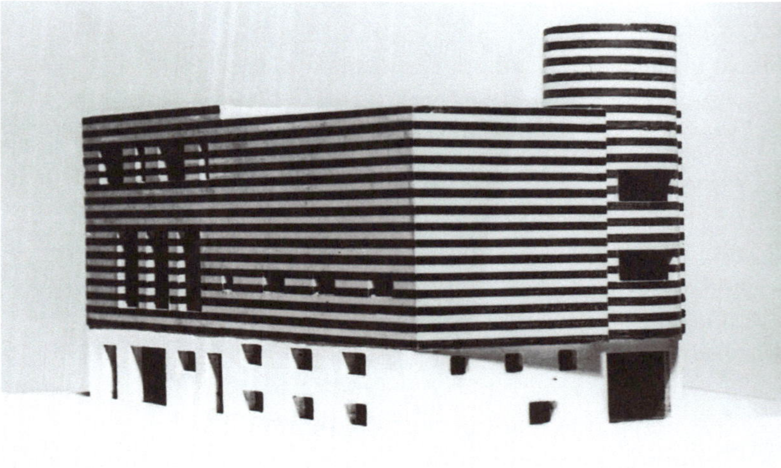

Img. 1

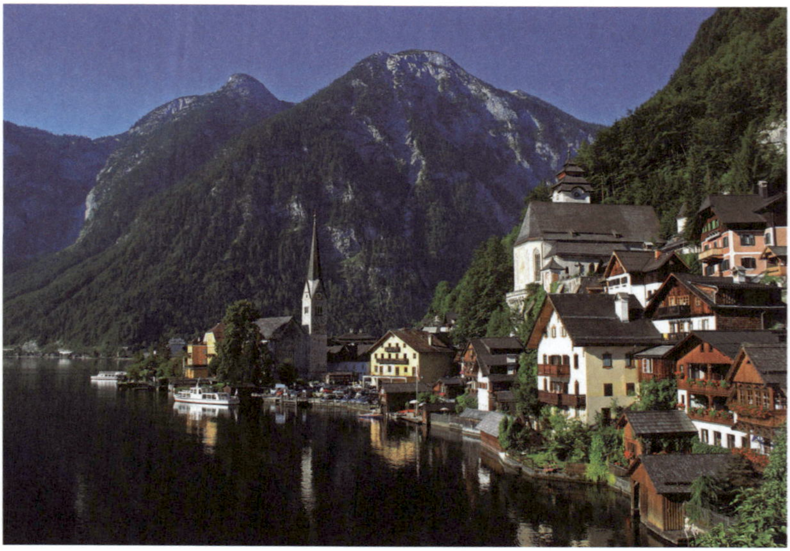

Img. 2

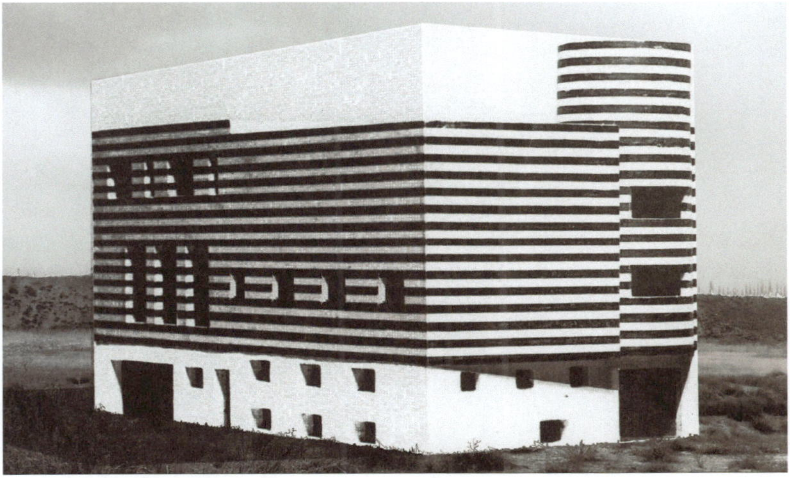

Img. 3

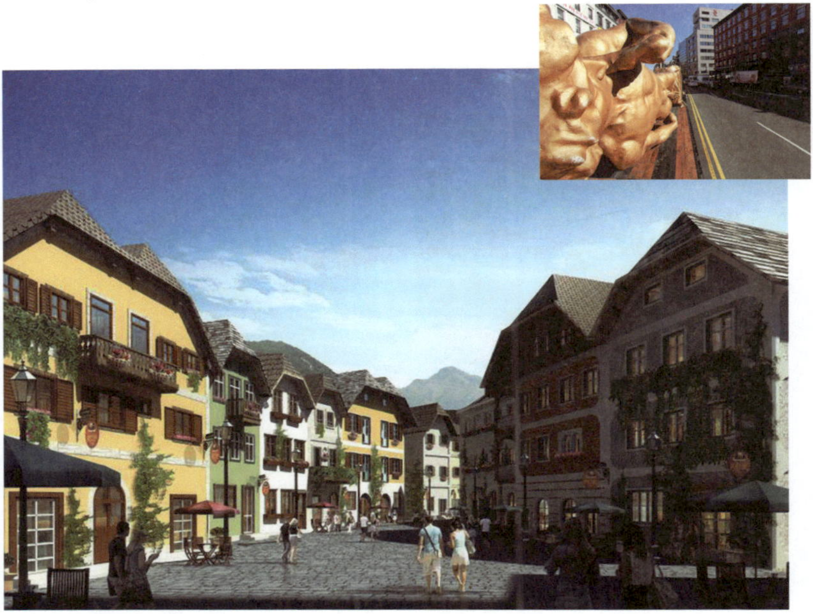

Img. 4

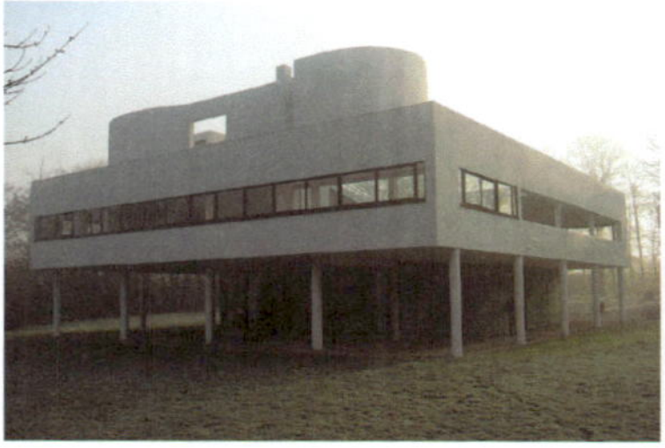

Img. 5-1

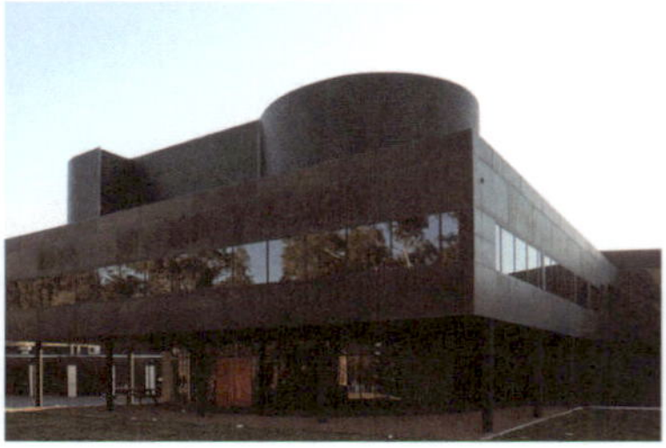

Img. 5-2

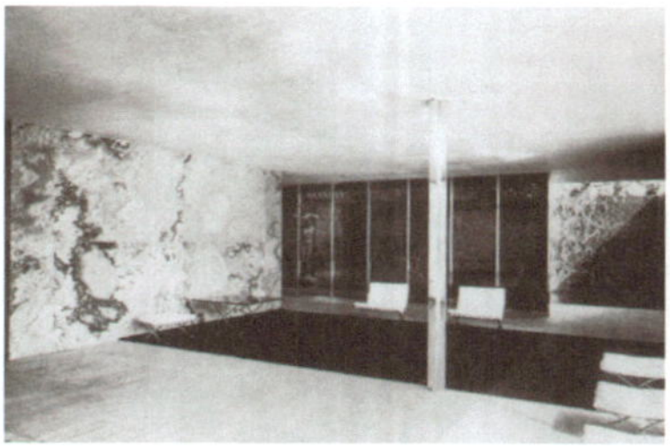

Img. 6-1

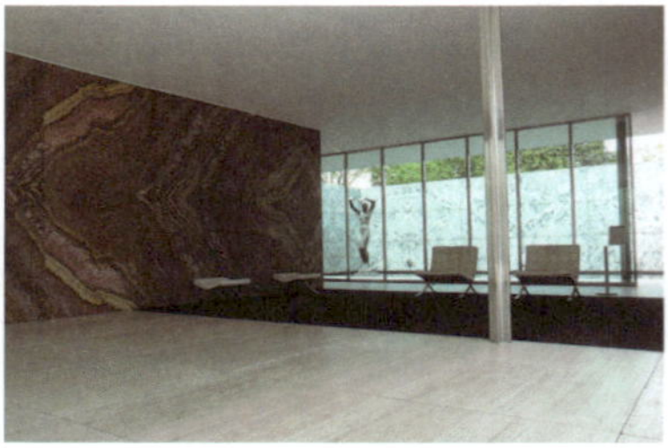

Img. 6-2

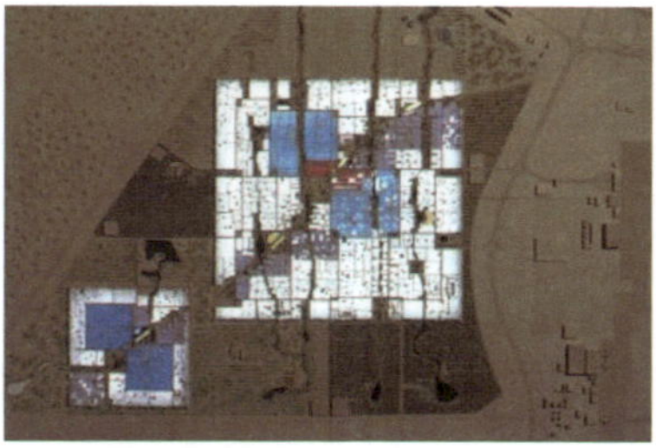

Img. 7-1

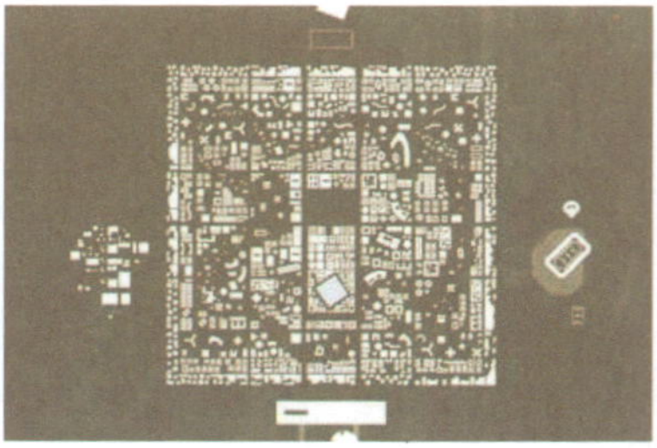

Img. 7-2

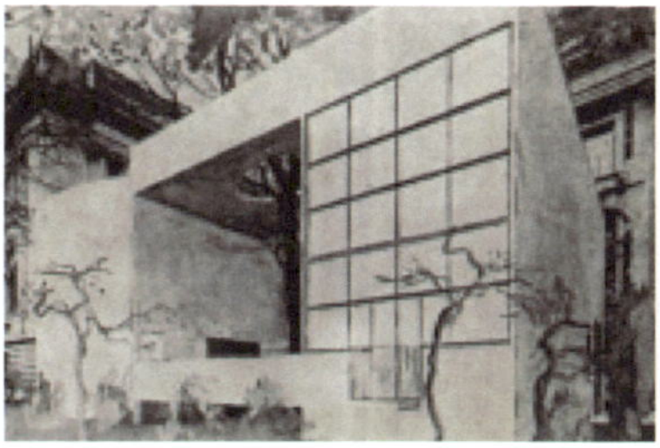

Img. 8-1

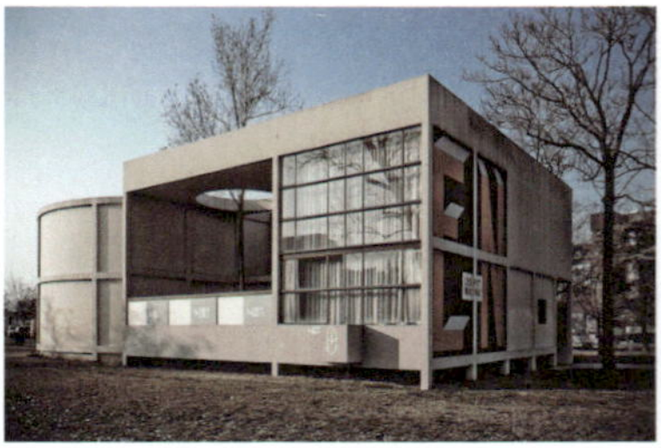

Img. 8-2

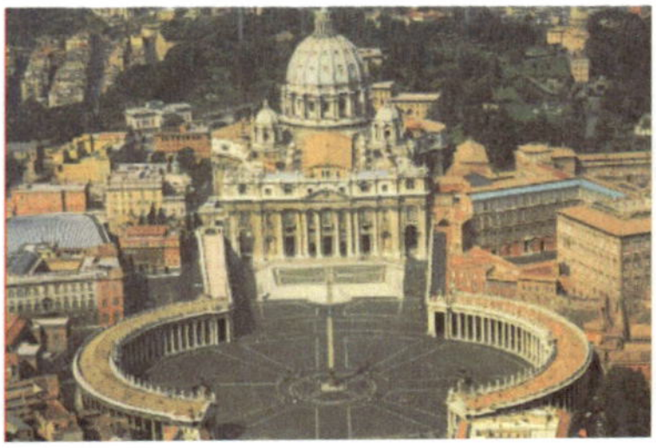

Img. 9-1

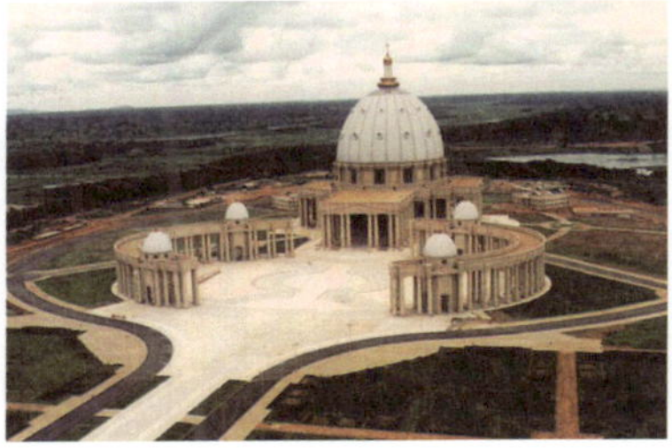

Img. 9-2

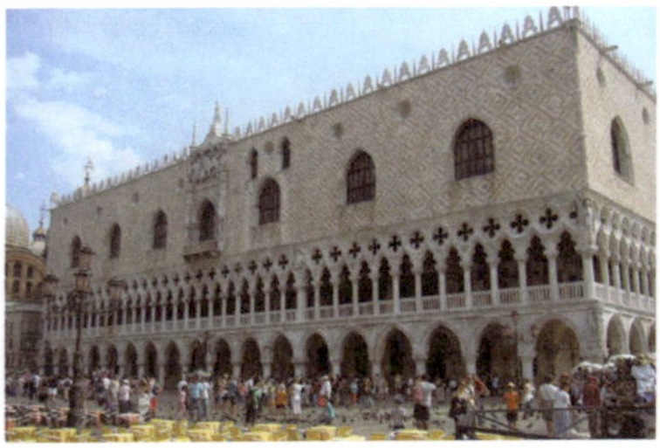

Img. 10-1

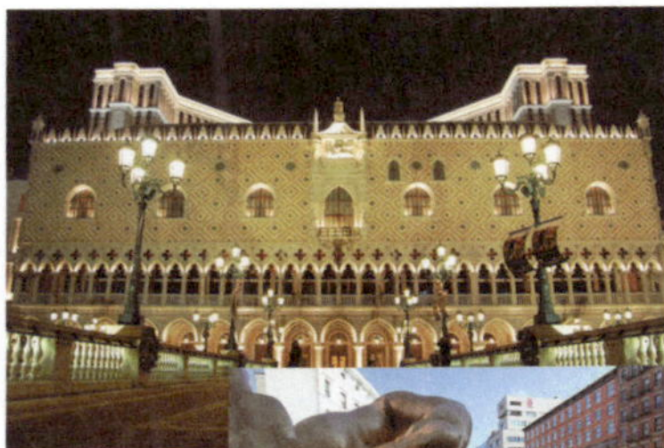

Img. 10-2

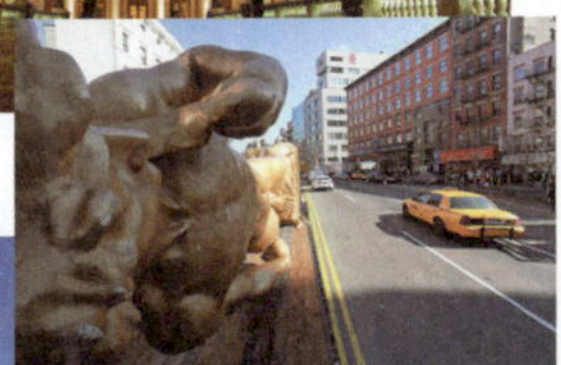

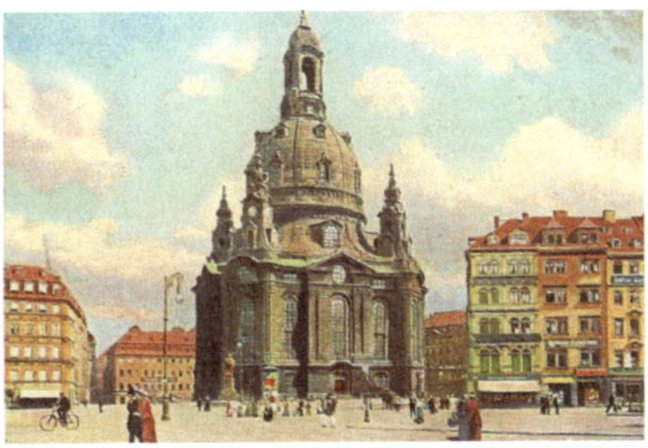

Img. 11-1

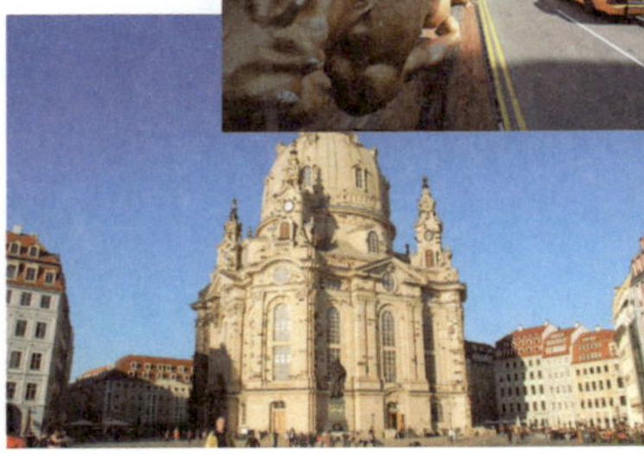

Img. 11-2

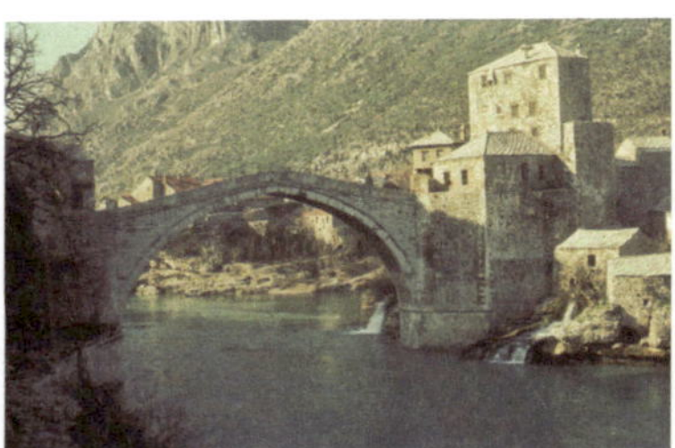

Img. 12-1

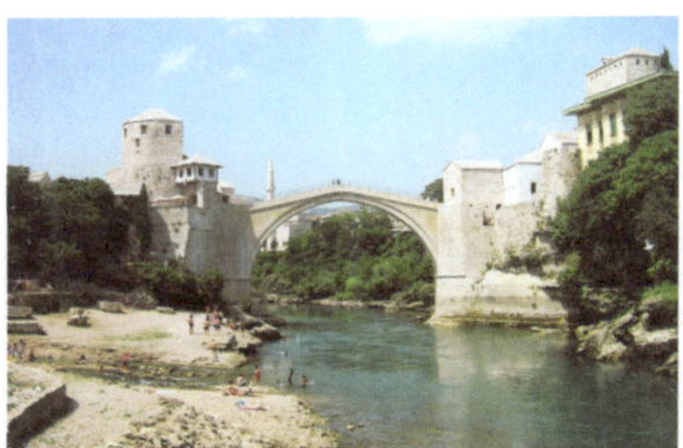

Img. 12-2

Acknowledgments

This text developed over various encounters with doppelgangers, architectural or human ones, in reality and in fiction. It also took shape in the work together with Sam Jacob in the research cluster "Architectural Doppelgangers" that we are currently directing at the AA. I would like to thank Sam, and his and my students for their many beautiful ideas and arguments. Also, in the tradition of copyright liberation, this text is partially the result of my rummaging through Eyal's. The liability, of course, is all mine (and the text's).

DOUBLE

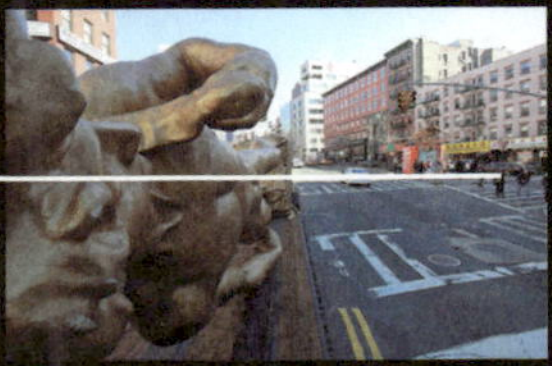

Fake Industries
Architectural Agonism

In a time when the bursting of the real estate bubble has lowered the rate of new construction; when a credit crisis hinders funding for building; when rampant foreclosures render useless more design; when architecture's intelligentsia delve into preservation as an attempt to grow with or without expansion; when global civilian revolts reclaim the right for dissensus; when the only role that architecture plays in the current European revolution is to burn; when architecture is absent in all "Springs" (in the Arab Spring, in la Primavera Valenciana); when the Indignados movement heartily distrusts architecture; when the closest that architecture gets to Occupy Wall Street is through hosting banks, which busy themselves by emptying architecture in the nightmare of foreclosures; in this time, we argue that the only way to proceed is through Agonistic Replicas.

The world is full of architecture—more or less interesting—and we do not want to add any more.

Instead, we advocate for the reconstruction of architectural imaginaries that had been forgotten, destroyed, overlooked, or dismissed. We propose the salvation of architecture in danger of extinction: the rebuilding of the hazardous architecture destroyed, as it was, for political reasons; the reimaging of this generation's domestic dreams annihilated, as they were, by the credit crisis; the rewriting of those passages of architectural history that have been tendentiously discarded.

In other words, we propose Agonistic Replicas, or Replicas in the sense denoted in the Roman Language: as a duplicate and as answer to a previous statement.

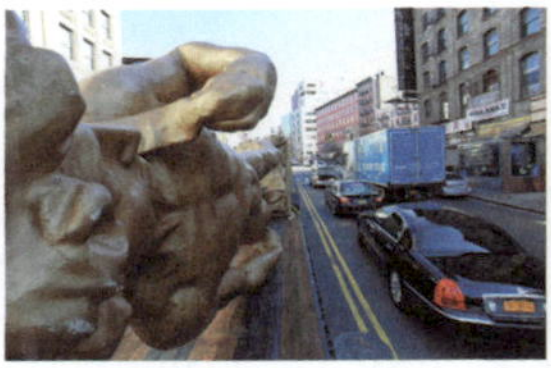

The unquestioned naturalization of certain issues within the field and the rapid destruction of others demands a reexamination that could lead to disagreement and therefore to disciplinary advances. The recent orgy of construction and destruction in the West calls for action: attack its roots, its claims for originality—copy, always double.

HOLY HOLY COPY COPY CULTURE CULTURE

A MANIFESTO

Hrag Vartanian

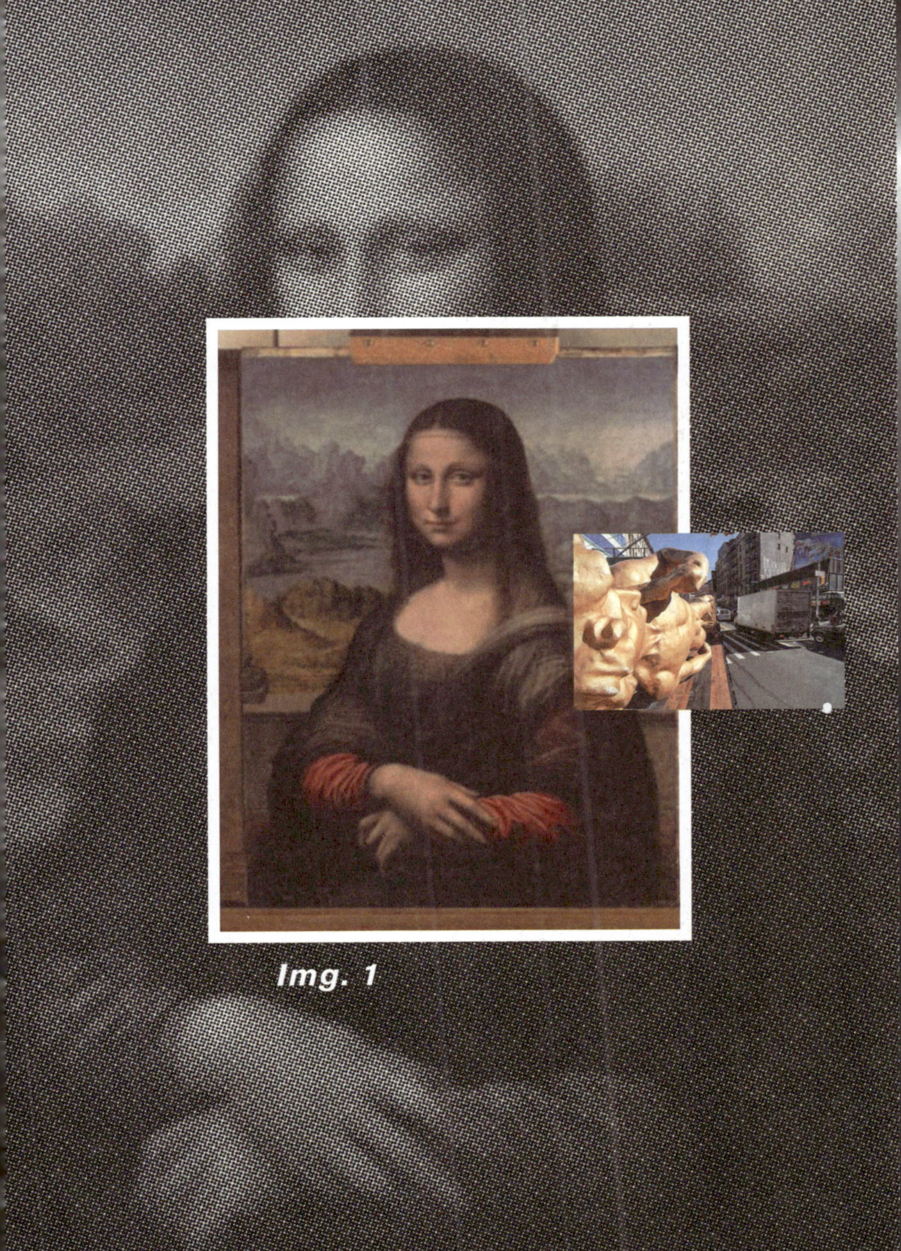

Img. 1

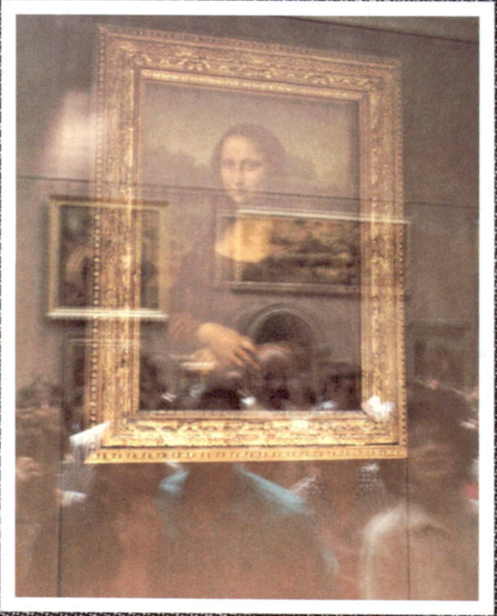

Img. 2

Everything in our culture can be copied and shared and "Liked."

Most (if not all) of us are unable to distinguish between the original and its offspring. Ownership has transferred from the physical object to the idea that it exists.

Something that is free and limitless, like an idea, can be copied ad nauseam, although not always legally.

When things can be infinitely reproduced they no longer have traditional value in their physical form. All art and objects are becoming conceptual.

This reality has made us all faith-based beings, even non-theists.

Our world is based on abstractions that we mutually agree upon and trade.

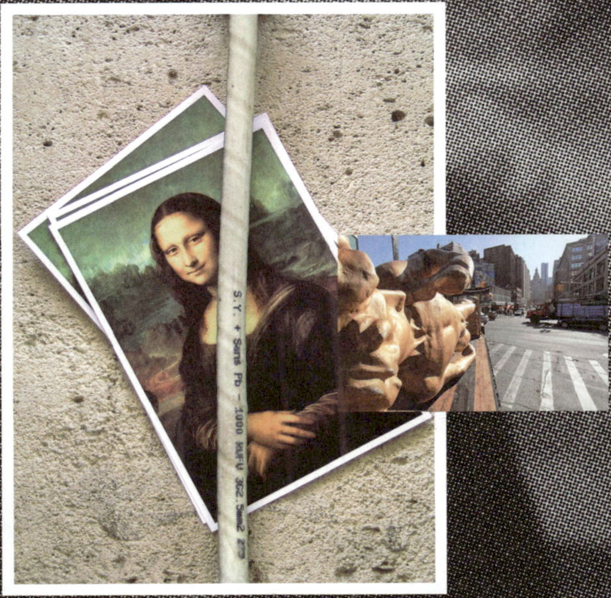

Img. 3

Copying acquired a negative association with the arrival of mass production, which made replication easy and cheap.

The cult of the individual artist meant that originality was prized above all else. If anyone could own it, then why desire it? But the aura of the original never disappeared and each copy made the source more desirable.

To own the original is to own the "mold."

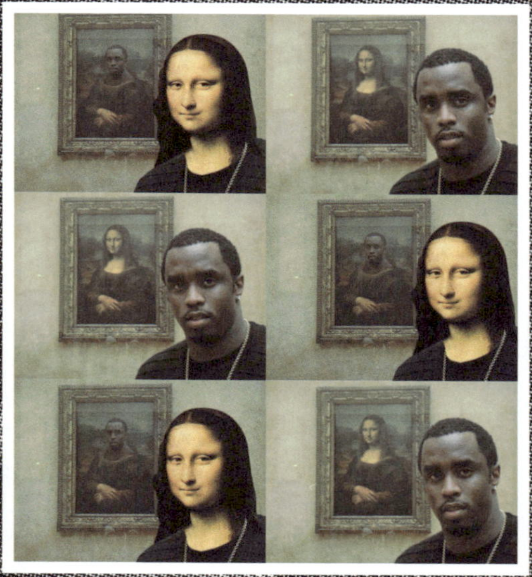

Img. 4

To see a copy is to know that an original exists, unless, as with an animated GIF, the reproduction is never meant to have a solitary original.

Proximity can be perceived as a form of knowledge. Being near beauty rubs off.

In the lyrics of American rapper Jay-Z, "uh, Picasso was alive he woulda made her / thats right nigga Mona Lisa can't fade her."

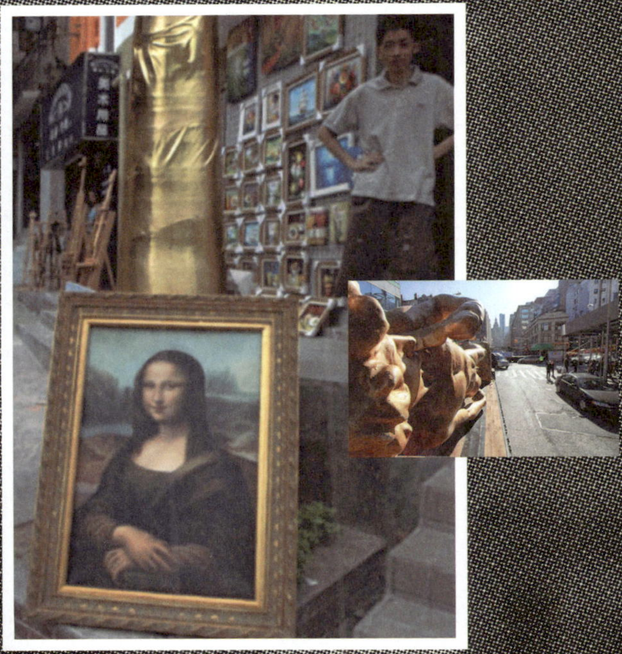

Img. 5

Original beauty surmounts imitators.

To want a copy is to want the original in another form.

To want a copy more than an original is not, however, uncommon. There is a responsibility in the original that many of us are not willing to take on.

To remake an image can be a form of meditation. To create it from scratch is a mini-miracle.

In the case of street art, the copy is irrelevant. Often the same thing is placed on the street and sold in galleries. Advertisement and original are one and the same. The copy is the message.

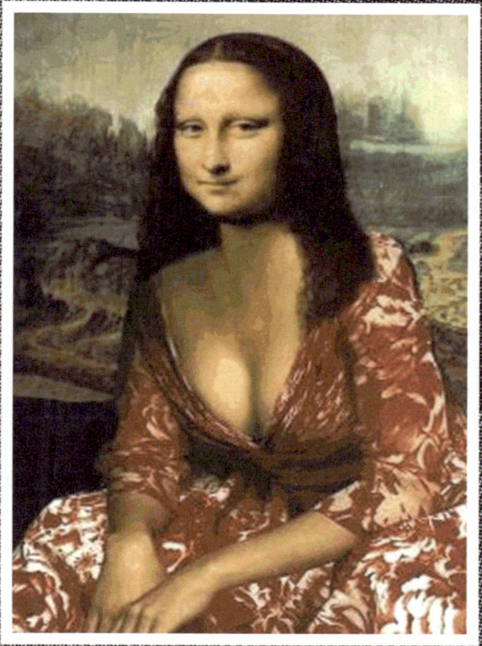

Img. 6

To remix an image is to celebrate it, to liberate it, to jostle it from its pedestal and breathe life into its carcass.

To take something apart is to understand it—like tinkering with a machine or looking through an archive.

Combining two or more things is an act of creation.

Nothing stands still.

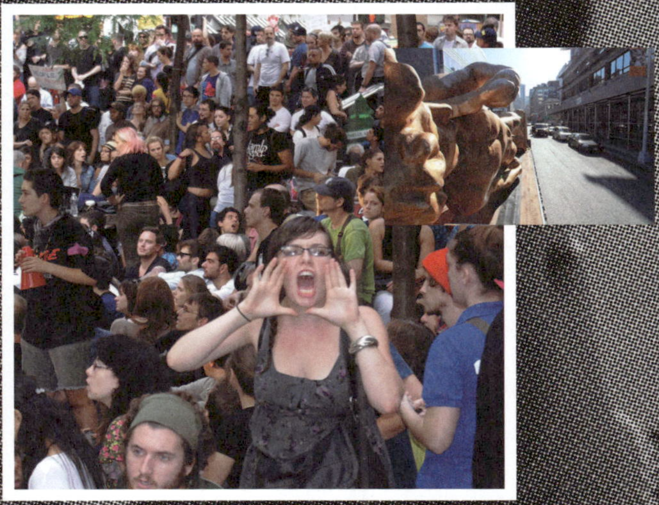

Img. 7

A copy is a form of echo. By reproducing something you become a conduit. The idea goes through you. As with the "people's mic" or a "reblog"there is a power in providing amplification to others. We take ownership in these actions, even if we shy away from being in the spotlight.

The most contemporary form of this idea is the meme, a work made to be copied. Researching memes can feel like a form of hagiography. Their symbolic currency surpasses any question of origins.

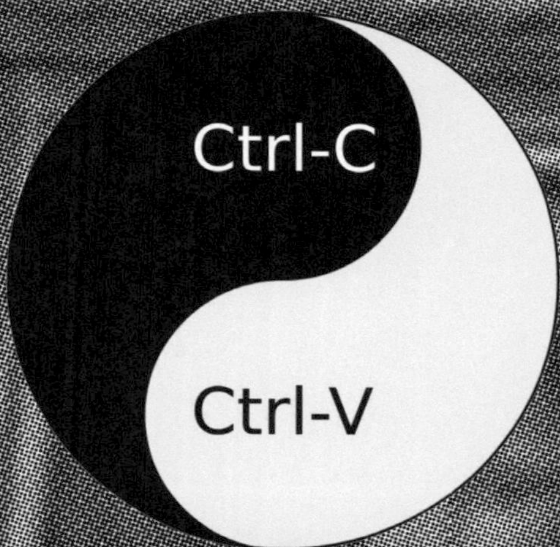

Img. 8

On January 5, 2012, Kopimism (pronounced Copy Me-ism) was

officially recognized as a religion in Sweden. It was propagated from a phrase in the Christian New Testament, 1 Corinthians 11:1, "Copy me, my brothers, just as I copy Christ himself."

In Kopimism, "Copy" and "Paste" are sacred symbols and adherents to the faith believe that knowledge should be accessible to all, and that the copying of information is ethically right.

The digital meme is meant to be copied. To argue for authenticity in a meme is absurd.

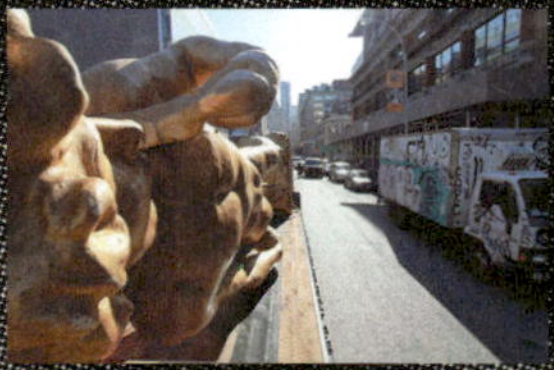

Citation is crucial to the copy. It guarantees the continuity of source and doesn't create a false notion of originality. A link in a chain, citation is a powerful tool that creates history.

Karl Marx said, "History repeats itself, first as tragedy, second as farce." I would add that in the third instance it becomes a meme. George Santayana wrote, "Those who cannot remember the past are condemned to repeat it," but remembering doesn't ensure originality and copying can be sacred.

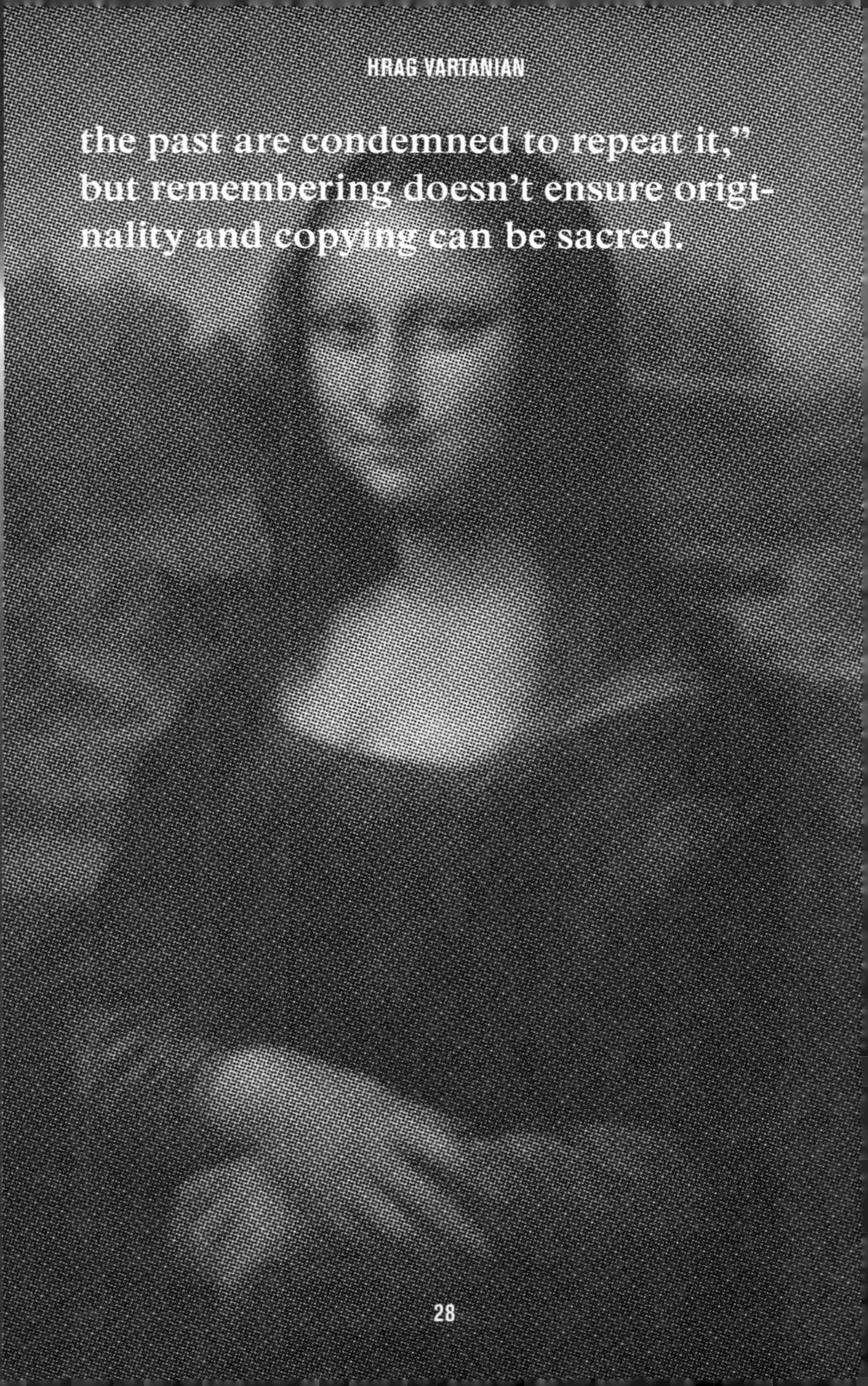

ANIMATED DOUBLES

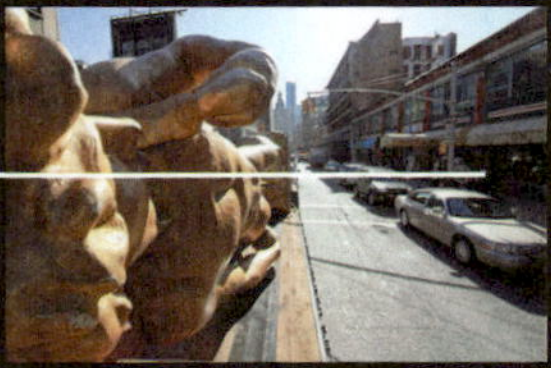

Spyros Papapetros

I.

What if the apparition of a double signaled the unexpected meeting between not just two, but hundreds of duplicate bodies, images, or things? In Serkan Özkaya's *A Sudden Gust of Wind*, throngs of identical sheets of paper twist and fold as if they violently surge from the ground and then fly in discord, but without the animosity habitually fermented during the reappearance of one's doppelganger (Img. 1). A vigorous, yet peaceful insurrection; a startling yet ultimately uneventful event: it is as if the standardized equanimity of these floating pieces of plain A4 or letter size copy paper tranquilizes their revolutionary momentum while maintaining the thrust of their quasi-supernatural state of levitation.

In a video-interview, Özkaya states that the memory which triggered *A Sudden Gust of Wind* was that of a "disaster" similar to the catastrophic events one reads of every day in newspapers: (gas) explosions, (building) collapses, or (car, train, or airplane) crashes—all accidents that presuppose a sudden propulsion of movement that comes violently to a standstill.[1] But in Özkaya's *A Sudden Gust of Wind*, such an episode is absolved from any tragic or traumatic content; his staged incident cancels the sadistic satisfaction that accompanies the reading of disastrous events in printed media. Here, the papers bear no news, no concrete information: the lack of textual weight rehearses the loss of bodily gravity lifting the papers into a floating condition.[2] The lifeless surfaces are animated by the synergy of physical weightlessness and discursive blankness. Instead of burdening the space around them,

1. See video interview of the artist with Georgia Kotretsos at Boots Contemporary Arts Space, St. Louis, MO: http://www.youtube.com/watch?v=7jksPtuM0N4

2. Özkaya has in fact completed an art book project based on hand-copying of daily newspapers. See Serkan Özkaya, *Today could be a Day of Historical Importance* (Louisville, KY: artwithoutwalls, 2010).

the clear sheets illuminate it with a self-bearable lightness and post-explosive effervescence. Not only the paper sheets, but also the gaps between them appear to double, triple, or infinitely multiply space without segmenting it, stirring the empty room with their partial rotations.

Here lies a subtle difference between Özkaya's installation and the artworks after which it is titled, namely Jeff Wall's photograph *A Sudden Gust of Wind (after Hokusai)* (1993) (Img. 2) and Katsushika Hokusai's well known print *Ejiri in Suruga Province (a sudden gust of wind)* (1830–33) (Img. 3).[3] In both of these homonymous precedents, the pieces of paper escaping from a briefcase or folder and blown away by the wind appear to carry some type of written or other visual information that is irretrievably lost. Such loss does not register in Özkaya's installation; the blank papers do not erase but anticipate meaning, signaling a pre- or post-linguistic event. In fact, in their complimentary gesticulations the papers appear to correspond with one another as if communicating among themselves.

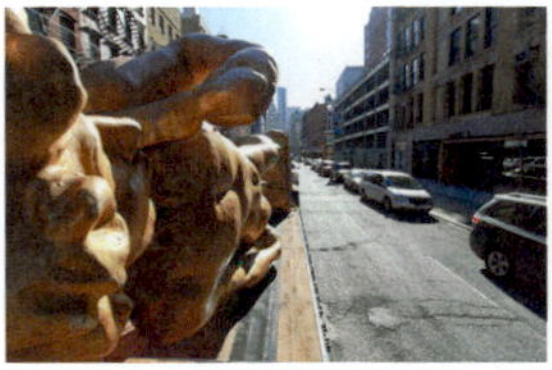

Even more striking is the absence of human bodies and any other signs of nature in Özkaya's installation. In Wall and Hokusai, humans appear entirely defenseless, physically and mentally overpowered by the unexpected vehemence of the natural current; while remaining precariously grounded, they too could ostensibly be swept away and disappear. Özkaya's installation rehearses the obliteration of human agency by

3. Hokusai's print was part of the original series of color woodblock prints published in Japan between 1831 and 1833 under the general title *Thirty-six Views of Mount Fuji*, the earliest series of landscape reproductions in the history of Japanese print.

its bodily absence; all expressions, such as surprise, fear, or resignation, are registered by the papers themselves that alternatively flatten or contort to convey revolt or acquiescence.

Yet the most staggering absence in Özkaya's *A Sudden Gust of Wind* is the loss of the main agent present in the title of all three works: the wind. Unlike Hokusai's print or Wall's photograph, both of which are staged in a natural setting, Özkaya's replication takes place indoors, which divulges the total absence of wind currents (this is valid for all [re]installations of the work, first in the Boots Contemporary Art Space in St. Louis [2008] and later in the Bilsar exhibition space in Istanbul [2009]).[4] The wind current becomes doubly (in)visible, uncannily phantasmatic. In Wall and Hokusai, the paper sheets fly in the same direction as the leaves blown away from a couple of trees so as to allegorize the parallel (under)currents traversing culture and nature. But Özkaya's installation does not copy, follow, or represent nature; his paper sheets appear to move in a direction that is self-willed, autonomous from any atmospheric or other natural condition.

There is a strong contrast between the buoyancy of Özkaya's paper stream and the solidity of the surrounding architecture, whether the exposed concrete and glass structure of Bilsar's industrial building in Istanbul or the white walls of the Boots Arts Space in St. Louis where the papers form a diagonal wave instead of a vertical pillar (Img. 4). In both spaces, architecture may appear to set a limit to the papers' levity, yet ultimately serves to reinforce it. Here, architecture is the wind and not the barrier that insulates air. Even if concrete

4. Versions of Özkaya's *A Sudden Gust of Wind* have also been installed in the exhibition spaces of *Cabinet* magazine in Brooklyn (2009) and DD172 media collective, New York (2010).

beams may metaphorically turn into trees and the striations of the wooden floor into miniature lakes or rivers, none of these architectural features framing Özkaya's installation has the plastic capacity for deformation that exists in nature. The papers themselves have acquired the rigidity of structural elements while simultaneously blossoming with a quasi-vegetal form of curly efflorescence. Suspended between petrifaction and movement, the paper surfaces are enlivened by a static form of animation. Indeed, in its obliteration of human and natural agency, Özkaya's *A Sudden Gust of Wind* portrays a form of life that is singularly invested in the object—an animation hermetically sealed from contextual or environmental factors. If the sculptural mobiles invented by Calder and other modernist artists were potentially activated by the movement of air, Özkaya's installation is immobilized by the presence of an artfully reconstructed vacuum.

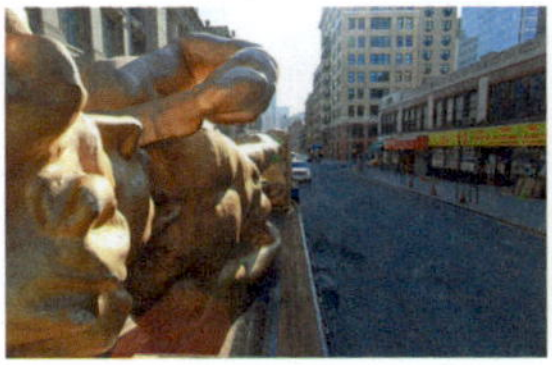

*

It is almost uncanny how literary and visual references to the double duplicate themselves endlessly in texts and images. It is as if duplication invites continuous replication until the presence or agency of the original copy entirely dissipates. Witness, for example, these two excerpts from literary fiction whose descriptions of the movement of tree leaves and paper pieces propelled by the wind appear to interleave with Wall's and Özkaya's copies. First is an excerpt from the 1915 novel

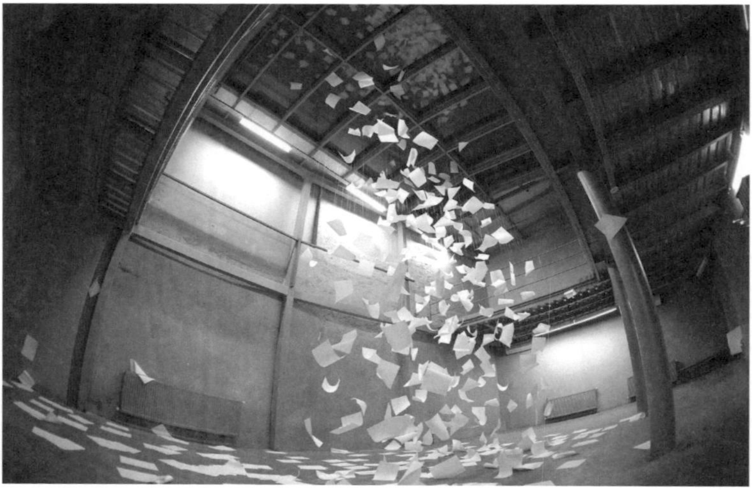

Img. 1

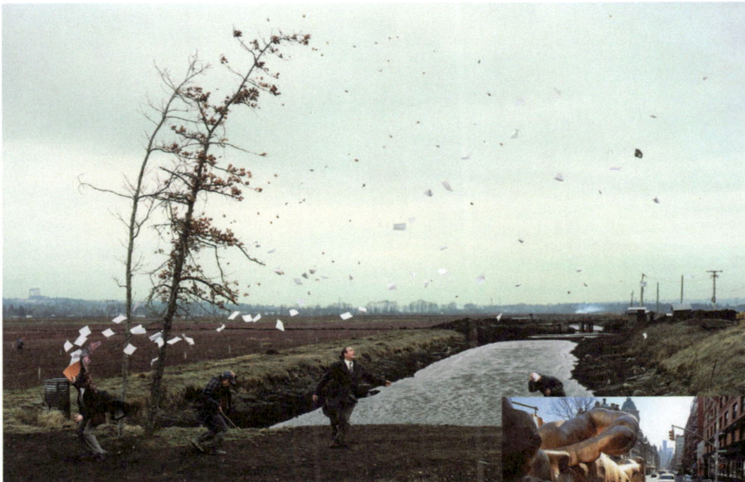

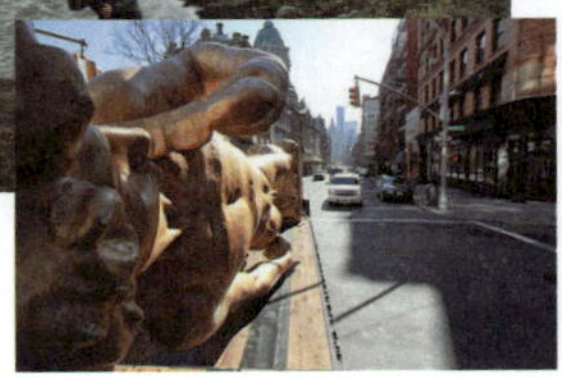

Img. 2

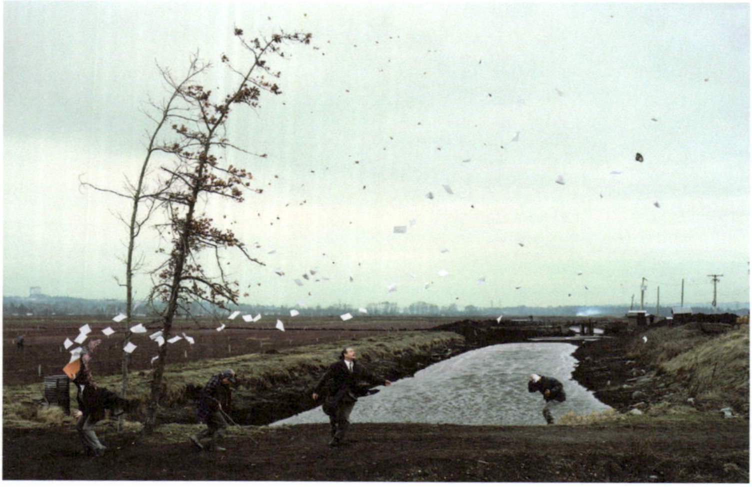

Img. 2

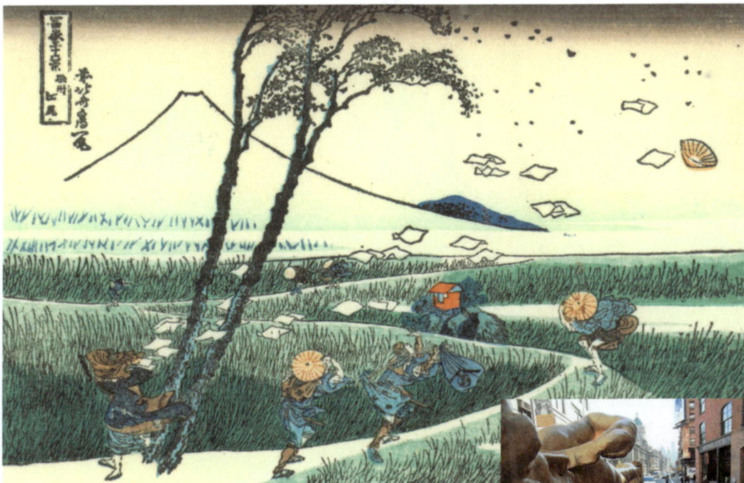

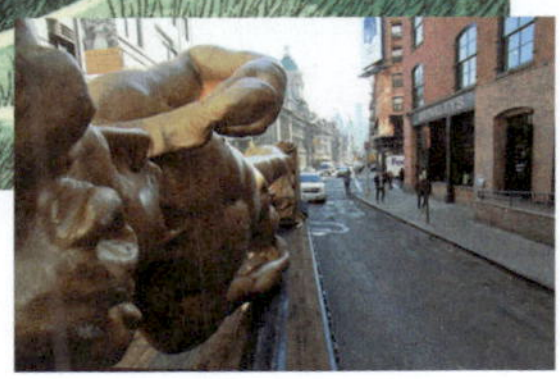

Img. 3

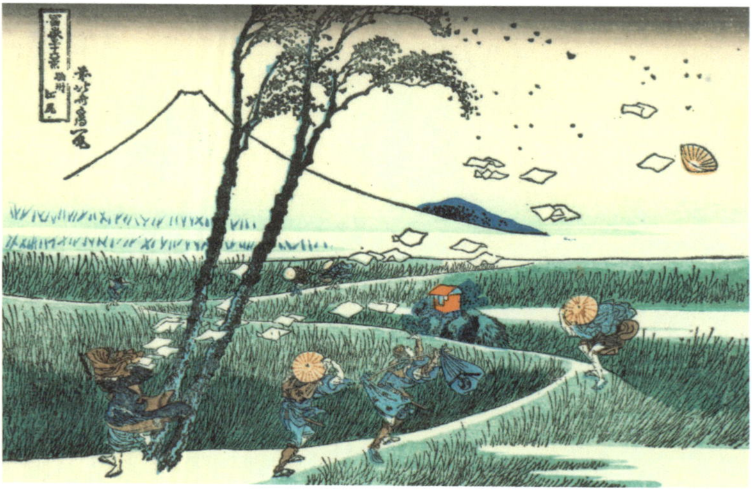

Img. 3

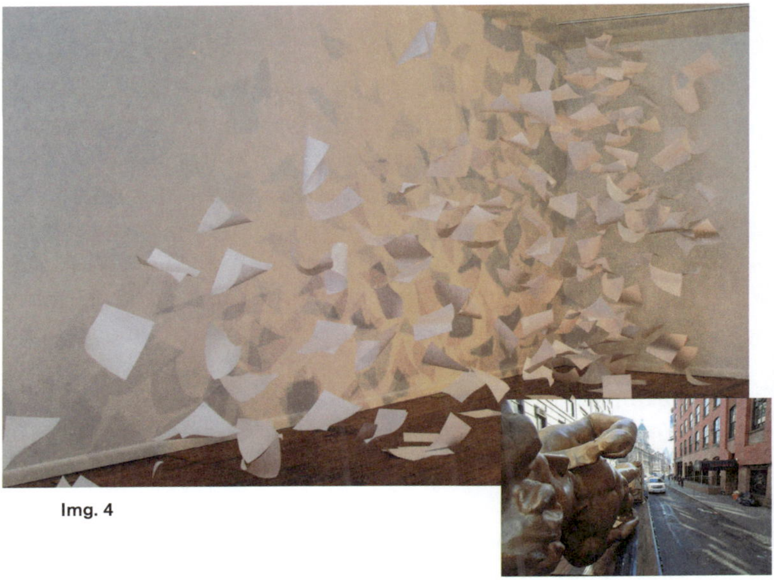

Img. 4

Der Golem, a contemporary retelling of the Jewish myth of a living statue restaged in turn-of-the-century Prague by the German expressionist author Gustav Meyrink:

> Once I stood in a desolate square and watched a whole heap of scraps of paper chasing one another. I couldn't feel the wind, as I was in the shelter of a house, but there they were, all chasing each other, murder in their hearts. Next instant they appeared to have decided on an armistice, but all of a sudden some unendurable puff of bitterness seemed to blow through the lot of them, and off they went again hounding on his next door neighbor till they disappeared round the corner. One heavy piece of newspaper only lagged behind; it lay helplessly on the pavement, flapping venomously up and down, like a fish out of water, gasping for air. I couldn't help the thought that rose in me: if we, when all's said and done, aren't something similar to these little bits of fluttering paper. Driven hither and thither by some invisible, incomprehensible 'wind' that dictates all our actions, while we in our simplicity we think we have free will. Supposing life really were nothing but that mysterious whirlwind of which the bible states, it "bloweth where it listeth, and thou hearest the sound thereof, but canst not tell whence it cometh and wither it goeth"![5]

As in Özkaya's *A Sudden Gust of Wind*, a memory reflex ignites the animation of inert paper surfaces. In a synergistic

5. Gustav Meyrink, *The Golem*, ed. and with an introduction by E. F. Bleiber (New York: Dover, 1976), 25.

form of empathy, the "flapping" of the curtain triggers the memory of the paper scraps' "armistice," and the gasping for air of a newspaper page recalls the biblical "whirlwind," which has a recognizable sound but an unknown origin. The passage on the wind that "bloweth where it listeth" is from the Gospel by Saint John (3.8), yet philologists argue that its origin, like any other textual or oral source, is not autonomous, as it echoes a number of ancient pagan cosmogonies describing the movement of air, spirit, or pneuma. In fact, Meyrink's own passage is a copy. Behind his memory of paper scraps chasing one another there is an aerial description extracted from a text much more recent than the Bible; namely Charles Dickens's *Martin Chuzzlewit*, a translation of which Meyrink had completed during the same period that he was writing *Der Golem*, and which includes the following scene:[6]

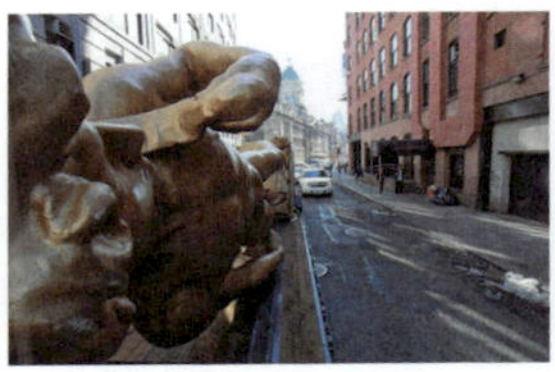

> **It was a small tyranny for a respectable wind to go wreaking its vengeance on such poor creatures as the fallen leaves, but this wind happening to come up with a great heap of them just after venting its humor on the insulted Dragon, and so disperse and scatter them that they fled away, pell-mell, some here, some there, rolling over each other, whirling round and round upon their edges, taking frantic flights into the air, and playing all manner of extraordinary gambols in the extremity of their distress. Nor was this enough for its malicious fury: for**

6. Bleiber, "Introduction" in Meyrink, *The Golem*, viii.

not content with driving them abroad, it changed small parties of them and hunted them into the saw-pit, and below the planks and timbers in the yard, and, scattering the sawdust in the air, it looked for them underneath, and when it did meet with any, whew! How it drew them on all followed at their heels![7]

There is an essential difference between Dickens's and Meyrink's descriptions concerning the portrayal of agency. In Dickens, the wind serves as the sole agent behind all actions—"wreaking," "hunting," "scattering," and "venting" like a dragon—while the fallen leaves appear as passive victims, entirely defenseless against the menace of the monstrous air. In Meyrink, the wind almost disappears, as the narrator states that he was inside "the shelter of a house" and "couldn't feel the wind." All agency is transferred into the scraps of newspaper, which appear autonomously motivated. In Meyrink, we transition from the symbolism of tree leaves to the codified information of printed matter. The agency of air is pocketed within the folds of the eloquent pieces of paper. Is it a mere coincidence that both Dickens's and Meyrink's novels were first published in serial form before being republished as books?[8] Can a literary description anticipate not only contextual factors but also the medium of its (re)publication?

While produced in very different contexts, critics have read the animated portrayals in Dickens's novel as reflections of early nineteenth-century angst against impeding

7. Charles Dickens, *Martin Chuzzlewit* (London: Penguin, 1978), 59.
8. Dickens's *Martin Chuzzlewit* was first published in nineteen monthly installments between 1843 and 1844. Meyrink's *Der Golem* was first published serially in the periodical *Die Weisen Blätter* from 1913 to 1914 (a fragment first appeared in *Pan* in 1911).
9. Stefanie Meier, *Animation and Mechanization in the novels of Charles Dickens* (Bern: Francke Verlag, 1982), 46; and Harold William Fawkner, *Animation and Reification in Dickens's Vision of the Life denying Society* (Upsala: University of Stockholm, 1977).

industrialization and the increasing mechanization of human life.[9] Even the violently dispersed papers of Hokusai's nineteenth-century print ostensibly replicate the aberrant proliferation of printed images between East and West in their increasingly global mechanical reproduction. Wall's creative replication has been read along similar lines, drawing additionally on the artist's digital manipulation of his images.[10] However, there is a major difference in temper among these visual and literary precedents and Özkaya's installation: his buoyant paper surfaces are not weighed by the animosity that pervades Meyrink's "murderous" scraps of newspaper or Dickens's wind Dragon. They are not persecuted by any vicious agent nor do they attempt to annihilate one another. They simply coexist in a non-hierarchical arrangement of mutual lightness. Their animation is sustained not by any external agent imposing on their movement but by the papers' own internal synergy—the heterogeneous, yet concerted expression of their shared substance and mutual contours.

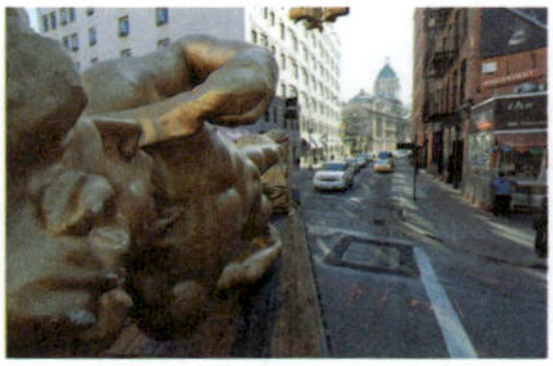

This unencumbered synergy in fact invokes preindustrial representations, such as the buoyant depictions of fabric "accessories in motion" in Renaissance painting and literature.[11] Even in paintings such as Botticelli's *Birth of Venus*, in which hair, fabrics, and small roses float, twist, and whirl, the presence of winds as allegorical personifications is superfluous, or as Michael Baxandall would argue,

10. See Laura Mulvey, "*A Sudden Gust of Wind (after Hokusai)*: from After to Before Photography," *Oxford Art Journal*, 30.1 (2007): 27–37.

11. The phrase "accessories-in-motion" *(bewegtes Beiwerk)* was coined by the art historian Aby Warburg in his dissertation on Botticelli (1892). See the chapter "The Movement of Accessories" in my *On the Animation of the Inorganic: Art, Architecture, and the Extension of Life* (Chicago: The University of Chicago Press, 2012), 30–69.

"eccentric"[12] (Img. 5)"It would be a good idea," states Alberti in his treatise *On Painting*, "when it is required to depict billowing garments... to portray the face of the winds... among the clouds blowing the garments."[13] The Renaissance author's use of the subjunctive suggests that the winds are in fact not necessary for the inanimate objects to start moving. The winds arrive after the accessories have already moved; they are no more than a rhetorical addition embellishing, but not propelling their movement. In other words, the "face" of the winds is another accessory, an accessory of the accessory, or an accessory of the second degree. Causes imitate effects in endless replication in the inverted epistemology of the animated image. Either personified as mythological deities, or hovering like invisible "sudden gusts," the winds reenter the scene as human masks only to announce their imminent departure.

In Özkaya's installation, the aerial current is replaced by quasi-invisible threads that hold all paper sheets together. As can be seen in the sunlight, thin connecting filaments made of translucent plastic (which Özkaya procured from a sock factory) are hung from the ceiling or anchored to walls and then attached to each piece of paper with the aid of an instantly drying spray glue (a type of "adhesive agent" only available in Turkey, apparently because of its lethal toxicity) (Img. 6). The geometric current of the spraying vapor is reminiscent of the luminous straight lines tracing the gust of air issuing from the mouth of Renaissance wind gods (Img. 7). Yet here, instead of propelling movement, the pneumatic substance instantly freezes it or kills it on the spot as if it were an insect. This

12. Michael Baxandall, *Giotto and the Orators: humanist observers of painting in Italy and the discovery of pictorial composition, 1350–1450* (Oxford: Clarendon Press, 1971), 138.

13. Leon Battista Alberti, *On Painting and On Sculpture: the Latin texts of De Pictura and De Statua*, ed. and trans. Cecil Grayson (London: Phaidon, 1972), 86–87.

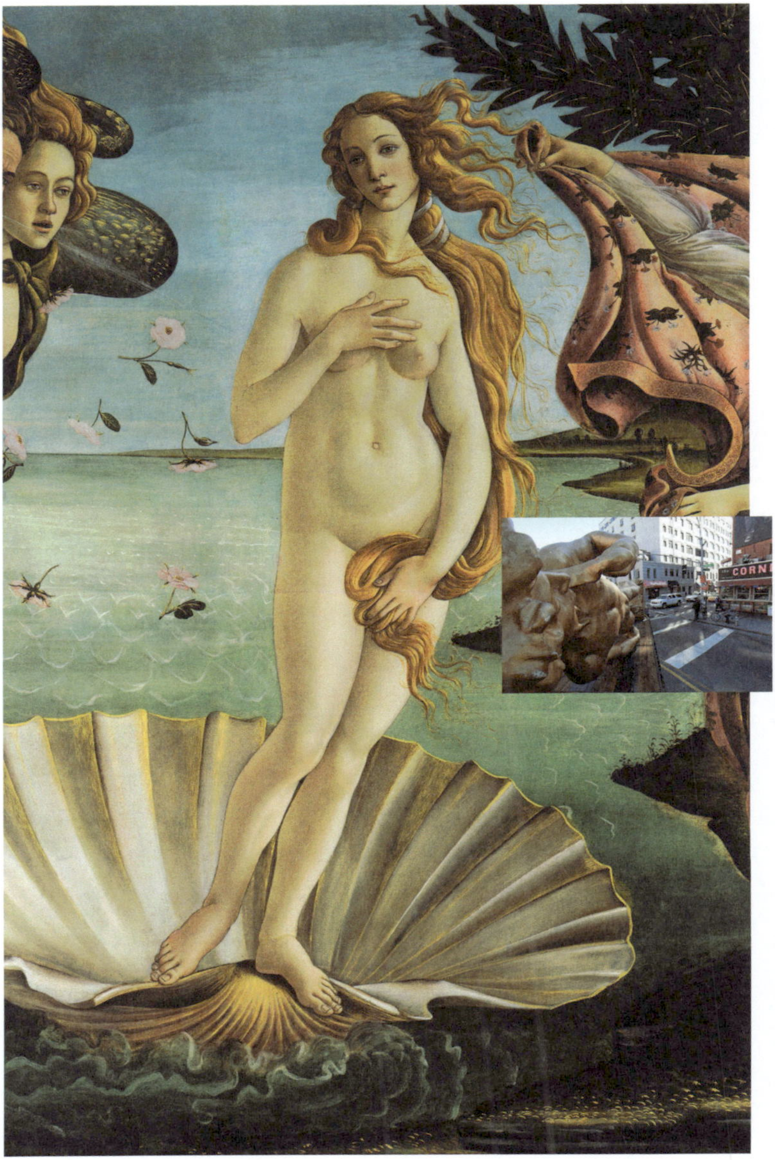

Img. 5

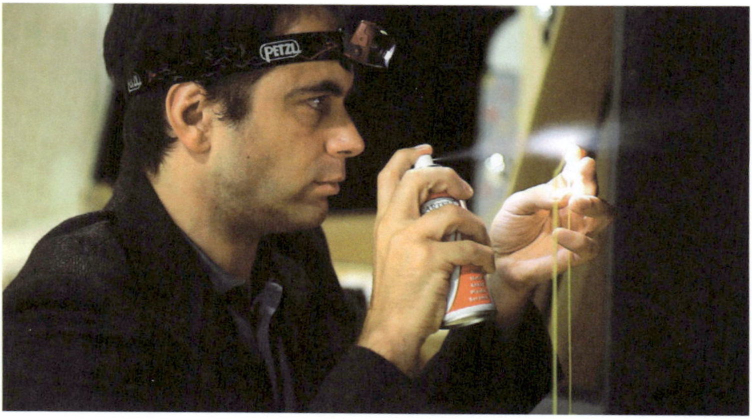

Img. 6

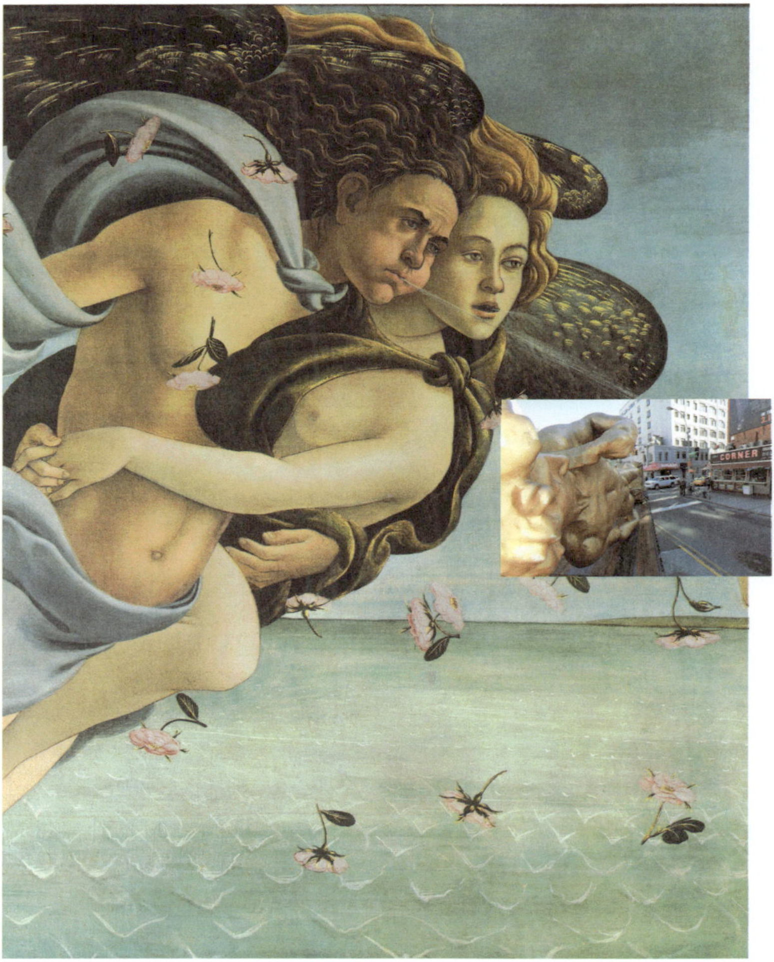

Img. 7

instant puff of adhesive agent is the only pneumatic trace in Özkaya's installation—the only real "gust of wind" and sign of aerial agency and life in his representation. Along with the synthetic fabric threads, the miniscule glue blotches act as invisible seams that turn the disparate crowd of papers into a fluttering, animated fabric.

II.

And now let us transition to a more direct reference to Renaissance iconography. However different they may appear, Özkaya's *Sudden Gust of Wind* and his replication of Michelangelo's *David* share an essential similarity based not only in the lightness of their material structures, but also in the erasure of authorial agency that characterizes both projects. In *Sudden Gust of Wind*, it was precisely the absence of the wind and its replacement by a network of invisible threads that animated the pile of papers in a manner that was almost uncanny. In *David (inspired by Michelangelo)* the lack of agency lies in the digital process that generated the flat layers of Styrofoam that built up Özkaya's original copy of the statue. The Turkish artist did not invent the program script used to scan and create a 3-D model of Michelangelo's sculpture.[14] His *David* is essentially neither his nor Michelangelo's. Instead, the statue becomes part of two collective processes: first, collective memory, the intersubjective process that accumulates infinite images of the same art object; and second, digital fabrication, the computer program that allows all surfaces of a three dimensional object to be reproduced in minute

14. The program was designed by Dr. Marc Levoy at Stanford University. Detailed information can be found at: http://graphics.stanford.edu/projects/mich/

detail. Özkaya's statue after Michelangelo is essentially a double fabric, as its own skin originates in material as well as cultural fabrication.

While (as critics have observed) the visual effects of air movement in Wall's *A Sudden Gust of Wind* were, at least partially, computer generated,[15] Özkaya's *David* is entirely (pre)fabricated by a digital program. Wall's synthetic image reminds us that from its origins photography was subject to minute darkroom manipulations, anticipating contemporary techniques of pixelation or other forms of digital air-brushing. Yet it is sculpture and not photography that today serves as a model for image reproduction due to computerized technology's predilection for 3-D modeling procedures. Özkaya's copies inadvertently create a new form of paragone, the Renaissance contest of the arts, here transformed into a (double) rivalry among sculpture and photography in terms of their adaptability for digital replication.

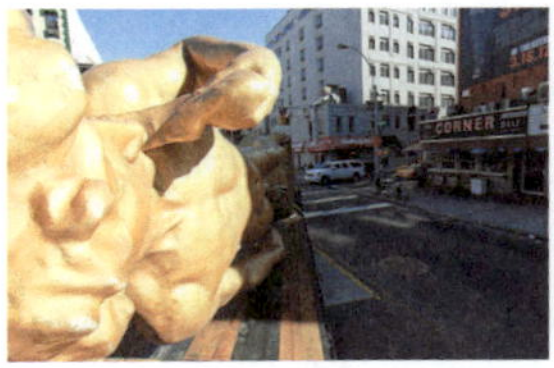

Özkaya's sculptural/digital copy of *David* is literally a double; he is double in height and eight times larger in mass in relation to Michelangelo's statue. It is said that Michelangelo used a single block of marble to model *David*; Özkaya uses hundreds of Styrofoam layers generated by a milling machine that are then glued together. A layer of gold paint—a glimpse perhaps of pop art and the master copyist Warhol—applied on the exterior does not entirely cover the layered core. Inside the thin coating one can still see

15. See Mulvey, "A Sudden Gust of Wind," 29–30; and Norman Bryson, "Jeff Wall: Enlightenment Boxes," *Art+Text* 56 (1997): 52.

traces of the several planes of foam, creating clusters of horizontal bands as in the geological planes of a landscape model.

Even if made out of foam, Özkaya's *David* is still deceptively heavy: it is as if the memory layer that envelopes Michelangelo's *David* in its original marble composition weighs more heavily on this replica than any of the more modern materials used in its assemblage. It was perhaps this illusory heaviness that misled the local workmen attempting to install the statue in a square in Istanbul into suspending themselves against the sculpture's frame, and which resulted in the collapse of two of the workers and the statue itself. Almost uncannily here, David replicated the fall of his opponent Goliath. Thankfully, as in Özkaya's *A Sudden Gust of Wind*, the disaster provoked was a relatively "light" one; no one was crushed by the fake gold or marble. Envision, however, the brief moment of suspension before the collapse, during which a living body precariously weighs itself against a human statue (Img. 8). This brief moment of balance is suddenly lost when the inert sculpture fails to provide the support illusorily transpired by its icon. While Özkaya's copy materializes the sculpture's self-replicating iconicity, its unintended fragmentation fulfills the statue's self-destructive potential: its fall, moreover, forewarns of a more general recession.

There is indeed something incomparably tragic in seeing statues collapsing from their pedestals. Seminal events in recent years (particularly in the Eastern part of the globe) have reconnected the collapse of statues to the falling of oppressive political regimes. In addition to his dominance in the art historical imaginary, Michelangelo's *David* is no stranger to discourses of political power. His inaugural placement was at the Piazza della Signoria, the very seat of social and economic authority in Florence. There is something in this combative, yet reflective young figure that compensates the establishment of power with its impending unseating.

In a photograph from the copy's initial assemblage in Istanbul in the summer of 2005, Özkaya's double *David* poses against an unfinished multiple story building intended for housing congress offices that here doubles instead as parking garage and artist's studio (Img. 9). The incomplete structure is adjacent to an army headquarters, from which (according to the artist), army officials would follow the daily progress of Özkaya's statue with binoculars. *David* returns—unceremoniously—at the seat of (Turkey's) political power as in his original installation in Florence. The federal building is the new Goliath which David now encounters. Against the exposed concrete, the magnified sculpture asserts not only its supersized scale, but also its foam-like lightness; the horizontal bands of the statue's foam layers measure in scale next to the building's stories.

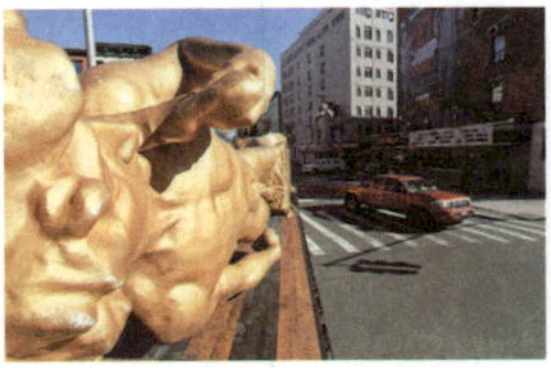

Özkaya's copy oscillates between 2-D and 3-D; it never is exactly either. Ultimately, the statue becomes a form of relief: a blown up image that bursts into the collective imaginary due to its very incompatibility with the surrounding architecture. In a still from the film *Double David* chronicling the adventures of Özkaya's copy in its world adventures since 2005, the director Gabriel Wrye places a model of Michelangelo's *David* against the Hagia Sophia, the most celebrated building in Istanbul, the city of origin of Özkaya's copy (Img. 10). Projected on the background of the Renaissance statue, the Byzantine church acts as a testament to both the contrasts and increasing similarities traversing Christianized Byzantium, repaganized Renaissance, and a contemporary Islamic country.[16] The photographic ensemble corroborates the preexisting allegiance between statue and building—the architectural and the sculptural monument—as they both frame and transform one another at a distance.

The sculptural copy's original fall from its plinth in Istanbul portends the replica's innate resistance to human manipulation. It is as if the statue itself defies permanent installation, which is congruent with the nature of a copy. Even the automobiles parked in the unfinished congress building used for its first assembly portend the statue's future relocations. Most web pictures depict Özkaya's *David* on the move: on top of a motor van, track, or cargo ship and strapped onto a solid frame as if the light structure were about to escape or fly of its own volition. After the statue's collapse, Özkaya had to essentially duplicate the broken replica by creating a mold made of fiberglass based on the original Styrofoam copy and then molding two other copies also made of (reinforced) fiberglass. During the four years (2006–2009) in which this new replication process was taking place in Eskisehir, the original copy was ultimately lost (after probably having been

16. The site intended for the statue's original installation during the 2005 Istanbul Biennale was in a different part of the city than the one shown in this film still – namely Sishane Square, another historical neighborhood in Istanbul with a number of mosques.

destroyed).[17] As in every mythological narrative, the object associated with its origin is necessarily lost so that its myth will survive. Özkaya's defunct double is survived by the two epigones sculpted after its original contour. Here "casting," a more traditional sculptural process, succeeds digital fabrication as a regressive reference to the origins of metal sculpture.

Michelangelo's *David* has been replicated a number of times. In addition to the one which resides in London's Victoria and Albert Museum, well known copies of Michelangelo's statue are to be found in front of a harbor in Copenhagen and in the observation point of the Piazzale Michelangelo outside the historical centre of Florence along with other copies of statues by the Italian master. Two years ago, an "official" copy of the Renaissance statue was exhibited for one day in the Piazza della Signoria, the location where Michelangelo's *David* stood for more than three and a half centuries before it moved indoors into the Gallery of the Accademia di Belle Arti in Florence. But in fact none of

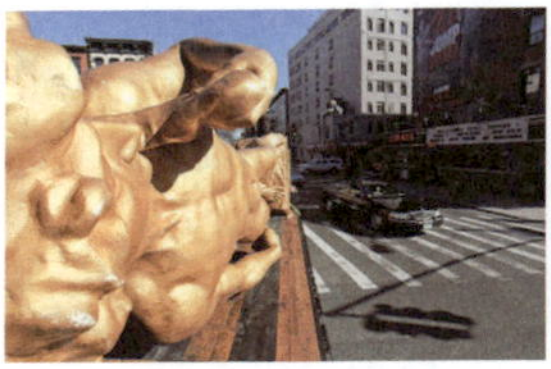

these locations was the one for which the sculpture was originally intended, that is, the rooftop of the Florence Duomo, the center of the city's religious power. As such, *David* is ostensibly displaced since birth; he never properly belongs to any place. Thanks to Özkaya he moves from Florence to Istanbul and to the streets of New York until he ends up in one of the busy streets of Louisville, Kentucky. While on the move, the statue turns from vertical to horizontal and thus transforms

17. The new project was sponsored by the municipality of Eskisehir and supported by the city's mayor who collects replicas from around the world. The mayor plans to install a second copy of Özkaya's *David* inside an artificial pond in Eskisehir.

from an immobile monument to a mobile object of festival pageantry (in which sculpture has historically played a central role). In some of the preliminary drawings for the New York parade he appears en profile facing the pedestrians; in others he lies flat on his back as if staring at the sky or the camera (Img. 11 – 12). As in *A Sudden Gust of Wind*, the main problem with this and every replication is orientation — the spatial direction copies follow when they circulate.

Architectural replicas, such as the Barcelona pavilion, are often tied to the place of their origin; the fabric and original dimensions can be altered during replication, but the location stays the same. Even when iconic buildings are geographically transported, as in the cases of the Nashville Parthenon or the endless replicas of Palladian villas in England and North America, such building replications ultimately become prominently local by being culturally and even nostalgically tied to their new location. In its seven year Odyssey across the globe, Özkaya's replica remained defiantly mobile until his latest relocation in front of the 21C Arts Center in Louisville Kentucky, where it appears to be treated with a sense of cultural ownership by locals (Img. 13).

*

Doubles create more doubles as copies by nature have a tendency to reproduce and even procreate. Objects are no longer autonomous in and of themselves but become part of sequential replication. Replication is their only authenticity. This does not only apply to the endless series of replicas of Michelangelo's *David*: two years after Özkaya first installed his *A Sudden Gust of Wind*, a British designer exhibited another installation of a similar assemblage of flying papers in the Victoria and Albert Museum in London (which, as mentioned, also houses the well-known copy of Michelangelo's *David*) in which curled paper sheets made of a more durable material are inscribed and sold individually as paper trays.[18] While this later installation looks strikingly similar to Özkaya's, the name of the Turkish artist was never mentioned. The case opens up a series of questions: Can one copyright a copy, and if so, what would the rights of copies or the copier be? Can copies ultimately turn into originals based on their inaugural reproduction? Are there degrees of authenticity that copies and originals can share? Such legal ramifications add an economic dimension to the double.

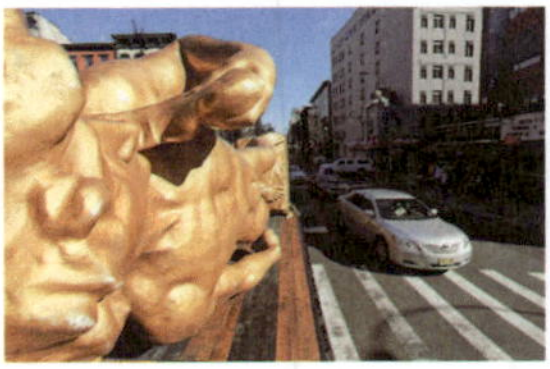

It is inevitable that doubles incite rivalry, whether on personal and psychological or artistic, cultural, social, and economic levels. In the 1913 film *The Student of Prague* by Stellan Rye and Paul Wegener (remade in 1926 and 1935 by other directors), which inspired Otto Rank's well-known psychoanalytic study of *The Double (Der Doppelgänger)*, a young man sells his soul to the devil and the devil substitutes the student's anima with a double of the young man's body that emerges from a mirror.[19]

18. *A Gust of Wind*: one-day installation by Paul Cocksedge Studio for the London Design Festival, V&A Museum (September 24, 2010).

19. Otto Rank, *The Double: A Psychoanalytic Study* (Chapel Hill, NC: The University of North Carolina Press, 2011).

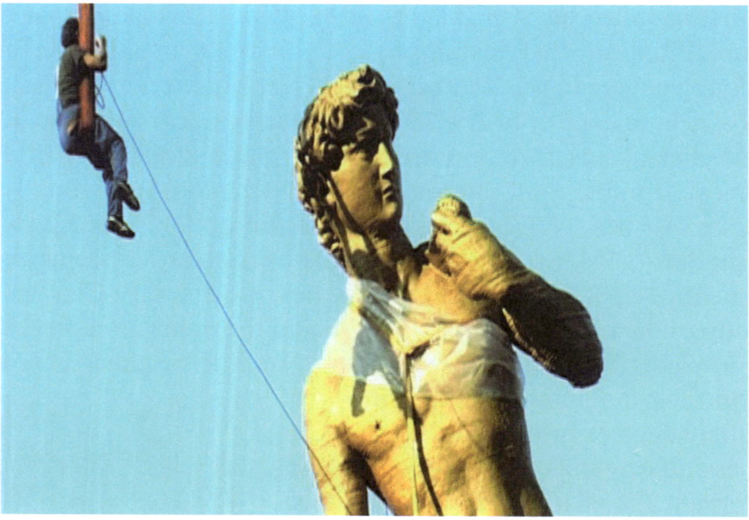

Img. 8

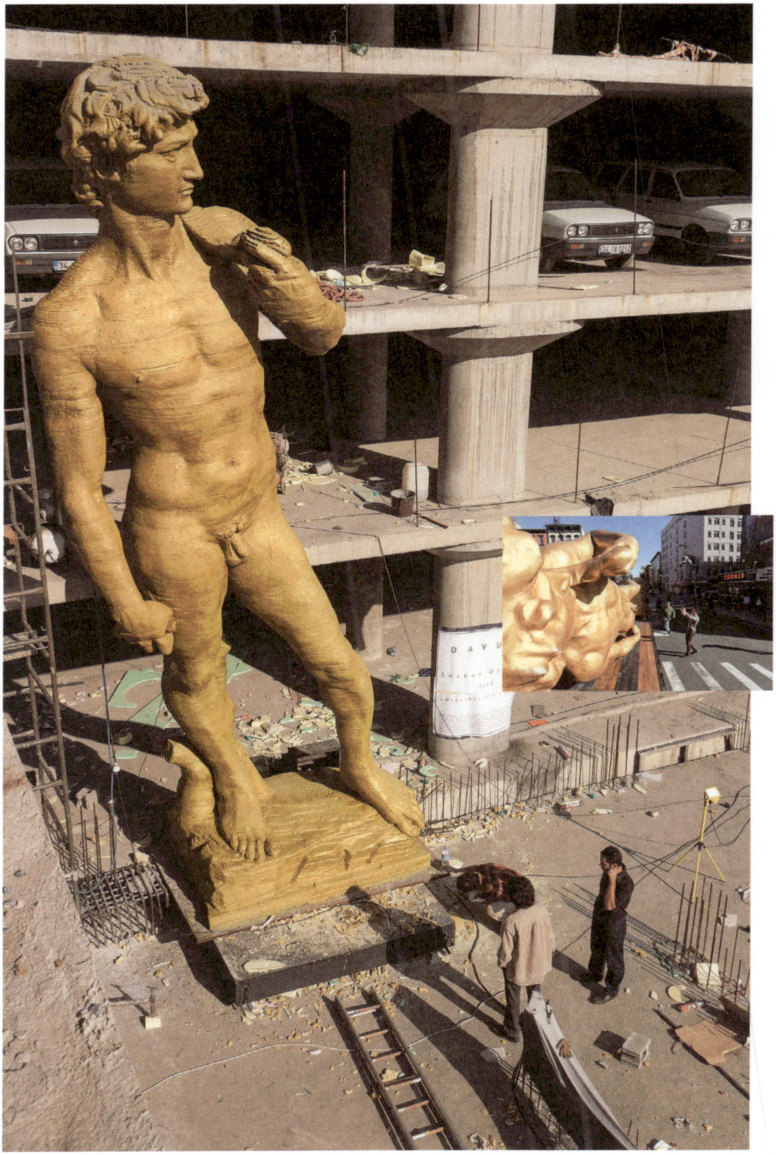

Img. 9

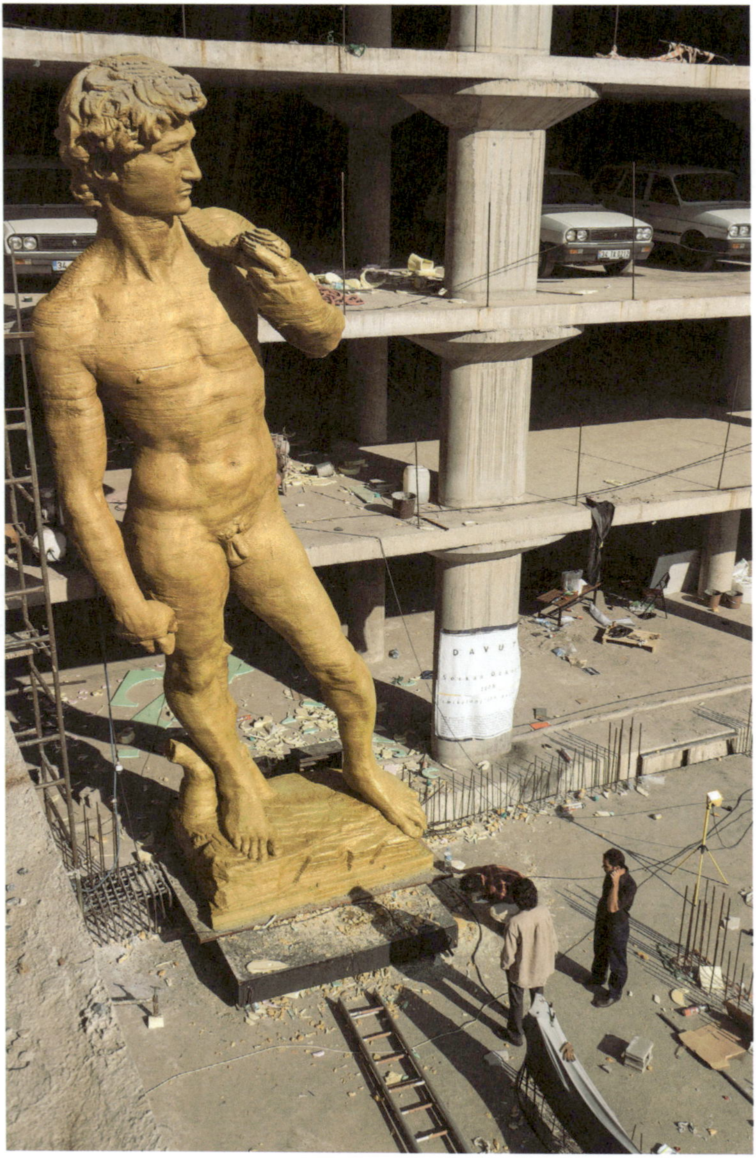

Img. 9

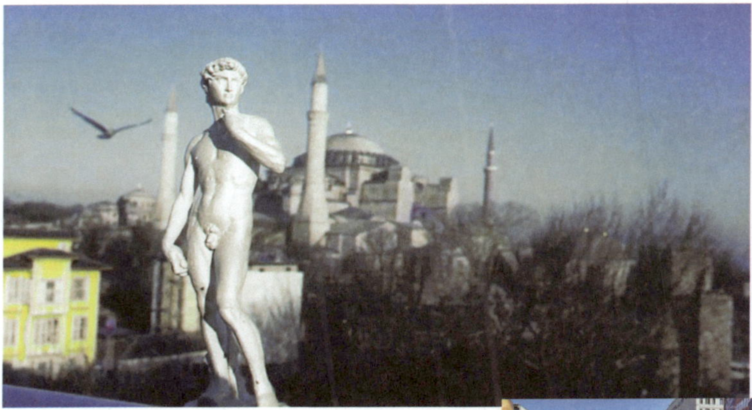

Img. 10

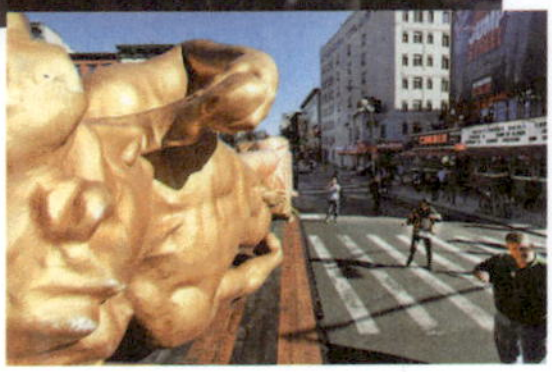

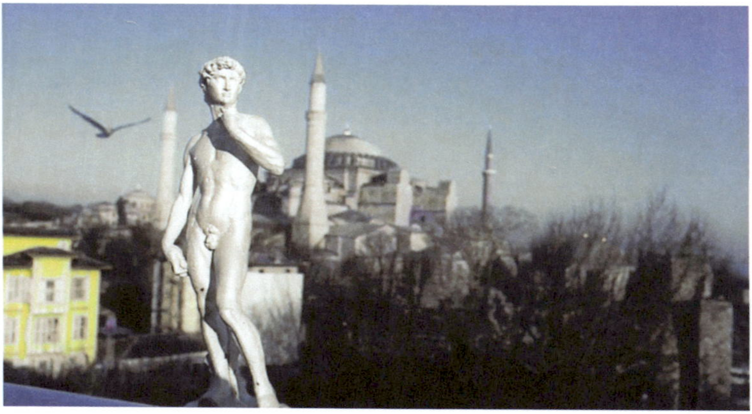

Img. 10

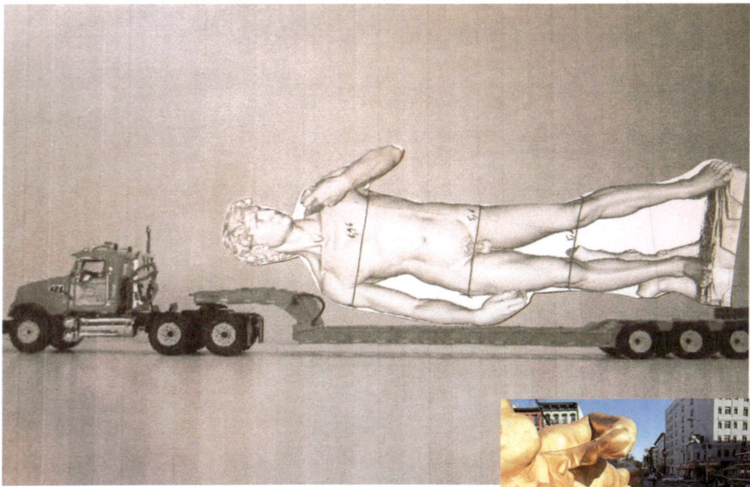

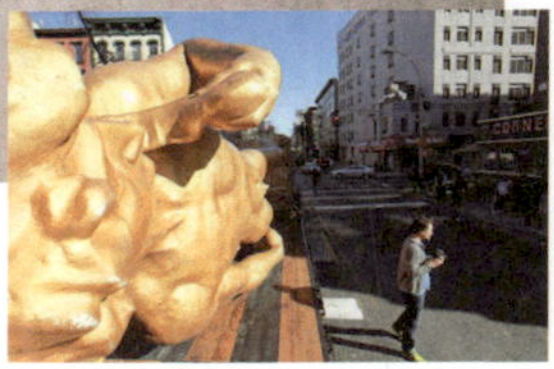

Img. 11

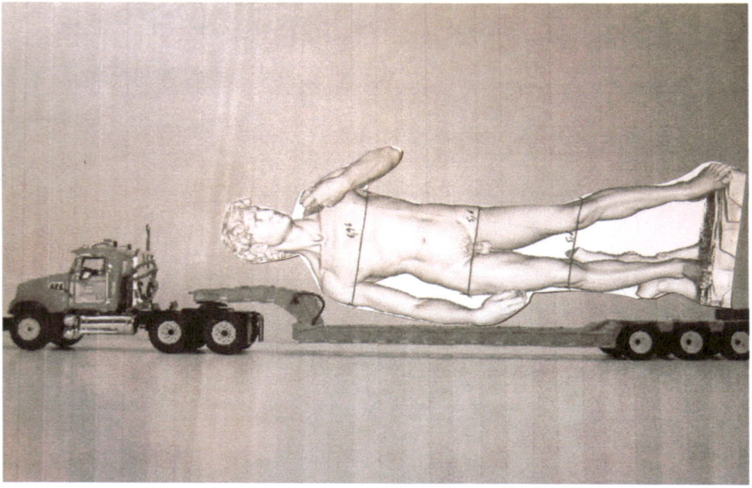

Img. 11

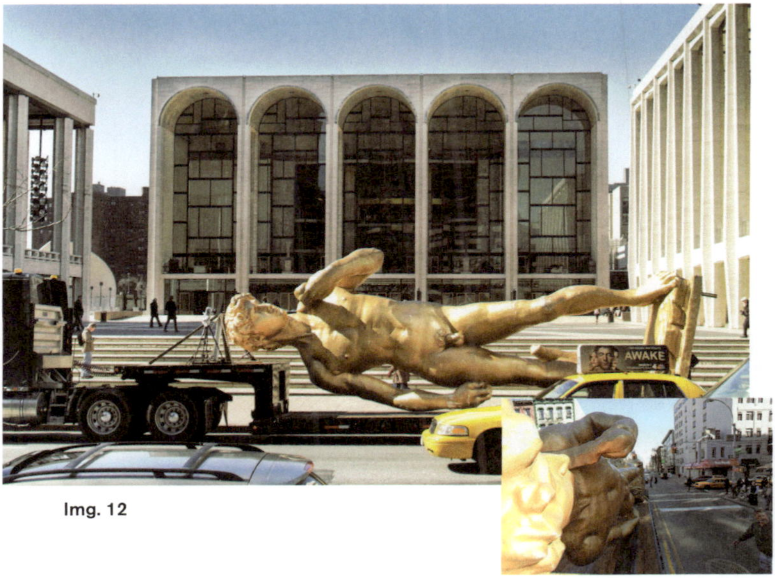

Img. 12

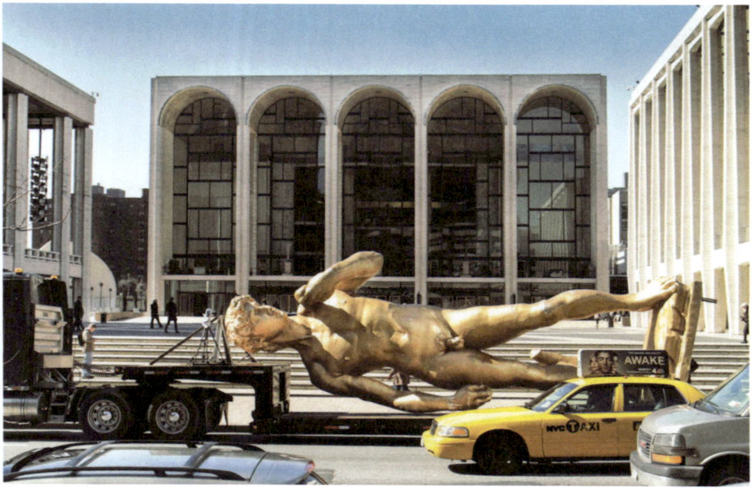

Img. 12

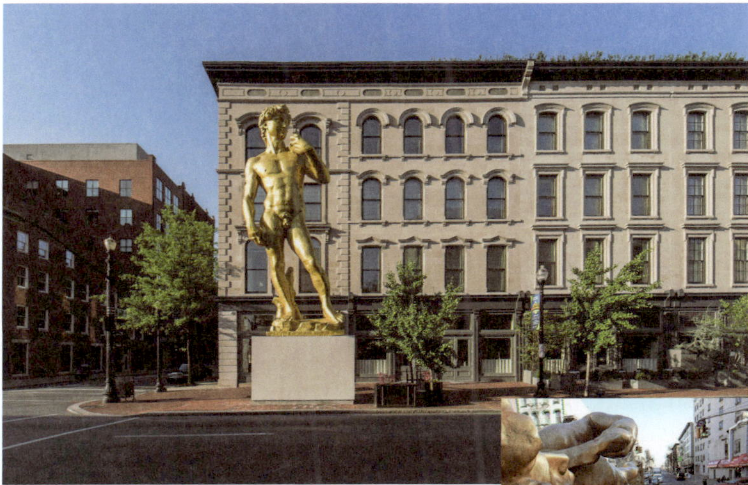

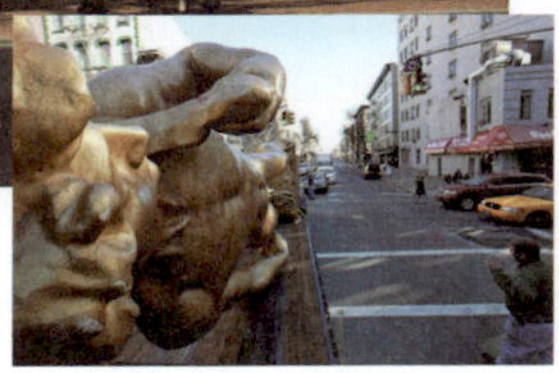

Img. 13

The very idea of a double has an ambivalent spatial dimension oscillating between projection and flatness. The double describes an image that springs forth from the bodily self and later moves against the original. In Rank's study, the double always appears menacing; it invites rivalry, constantly assailing the original self and even provoking him into a duel (Img. 14). In the finale of *The Student of Prague*, the young hero kills his double only to discover he has mortally wounded himself and expires immediately after. It is this tragic conclusion of the story that led Freud to declare in his 1919 essay "The Uncanny" that, having once been a product of "unbound self- love" and primary narcissism, the double "reverses its aspect" once this state has been surmounted and "[f]rom having been an assurance of immortality, it becomes the uncanny harbinger of death."[20] In the construction of artistic doubles, perhaps the "death" foretold by the psychoanalyst implies the demise not only of the original artwork but also that of each and every one of the copies in which the original both dissolves and proliferates.

The double is by nature architectural. A formal symmetry automatically monumentalizes both copy and original when viewed as an ensemble. Such monumentalizing symmetry is evident in *The Student of Prague*: there are always two female and two (or two by two) male figures: one young one old, one rich one poor, one happy one sad, one benevolent and the other malevolent and seeking to annihilate all others. This antithetical symmetry extends to the film's architecture: there are always two mirrors, two columns, two trees, or two

20. Sigmund Freud, "The Uncanny," in *The Standard Edition of the Complete Psychological Works of Sigmund Freud*, ed. and trans. James Strachey, Vol. 17 (London: Hogarth Press, 1953–74), 235.

candelabras that rhythmically punctuate the cinematic frame in every instance; thus, even when the young hero appears alone, the arrival of his double is expected. While the menacing "double" appears only sporadically for the greater part of the film, during the finale it shows up everywhere: the forest, the street, the student's room, and finally the cemetery. This omnipresence portends the ultimate perishing of both double and original when, following their final encounter, they both unite into a gravestone, the sculptural and/or figurative signifier of a memorial form of architecture. The gravestone is now the double of the dead hero and sculpture doubles as their mutually monumentalizing architecture.

What would then happen if *David* met the other *David* and if the two Davids started dueling with their slingshots? The meeting would reenact the David and Goliath story, but one could never be sure who was David or Goliath in such a reproduction. In Özkaya's copy, the two opponents are ostensibly commensurable. This retroactive res-

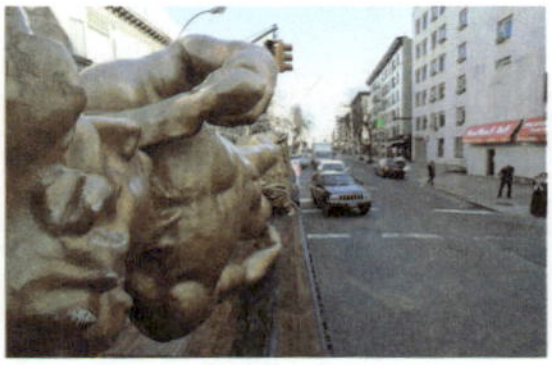

toration in size is part of the statue's projective or impending movement. Even if unequal in dimensions and weight, the two figures might ultimately compensate their differences before annihilating one another. Remember that David killed Goliath with only one shot from his self-made slingshot. Michelangelo allegedly portrays the moment of reflection in which *David* calculates this singular movement. The invisible projectile issued from the slingshot haunts the representation of the

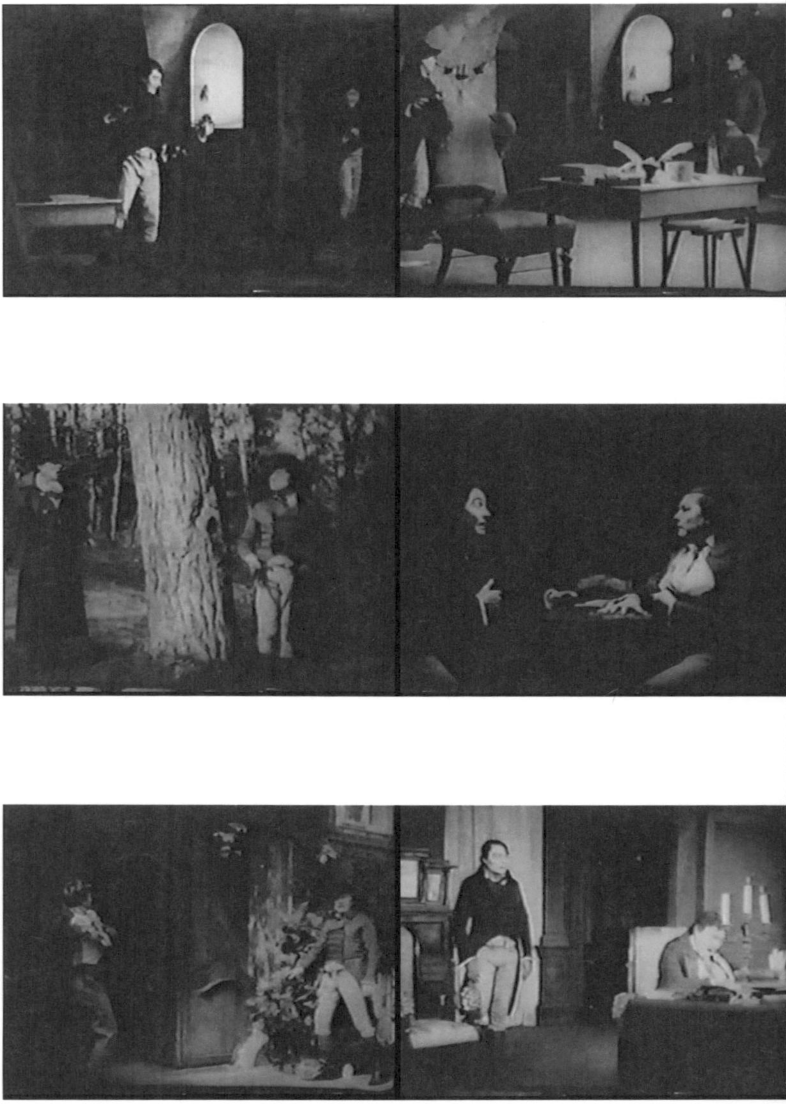

Img. 14

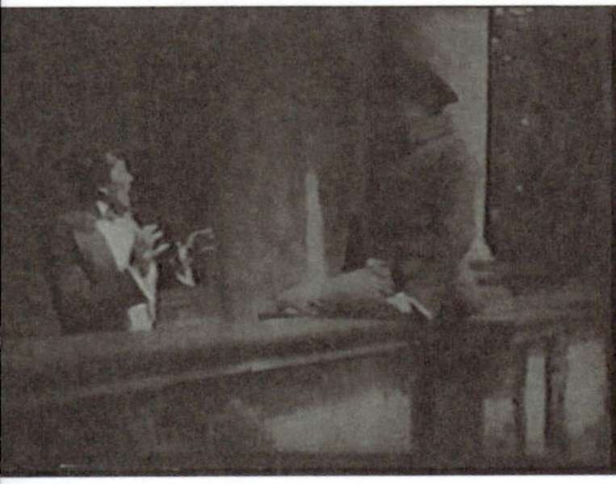

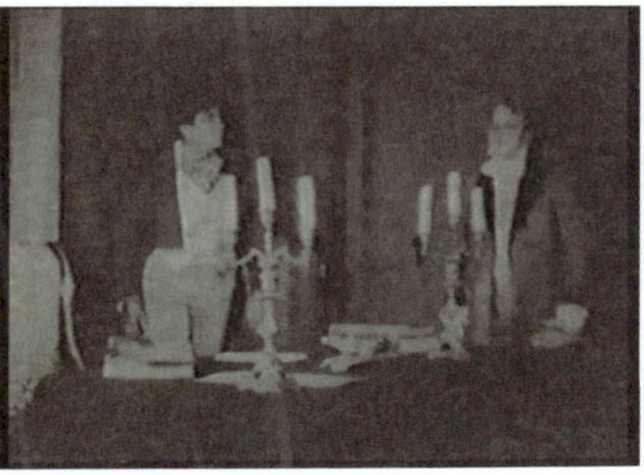

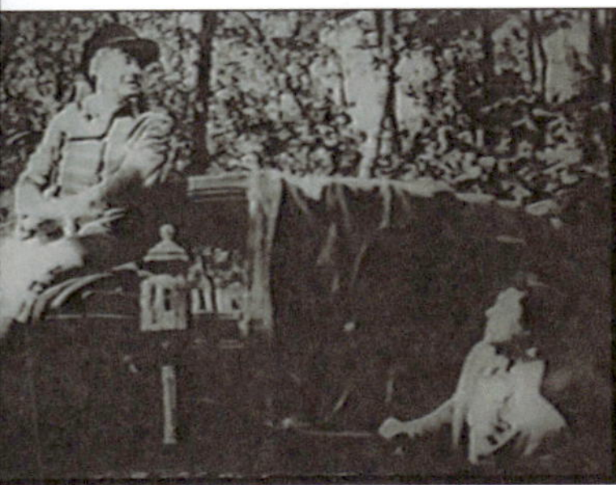

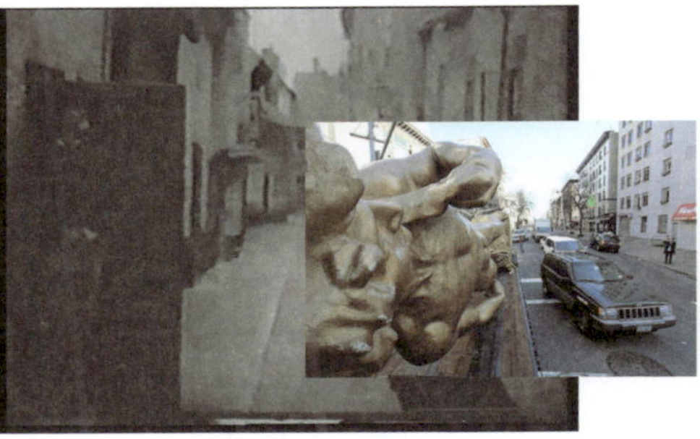

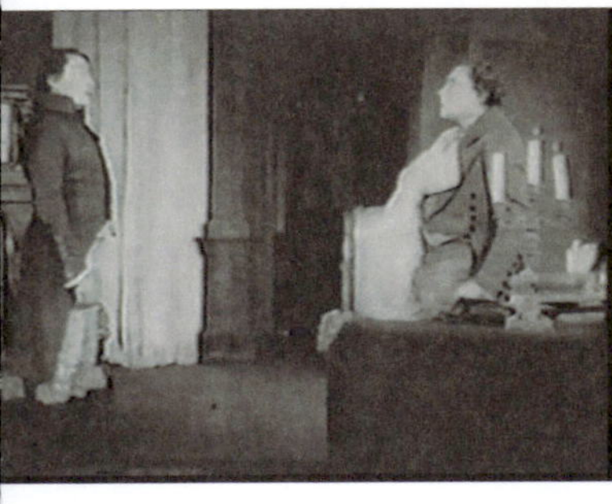

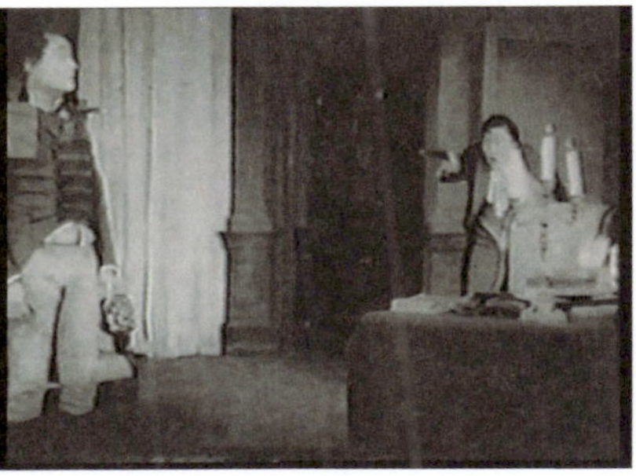

biblical hero. His entire body is a missile; an object caught at a moment of suspension. Özkaya's copy essentially replicates not equipoise but the arrest of movement. As analyzed by Freud in his detailed reading of Michelangelo's *Moses*, the sculptor freezes and eventually prolongs the moment that precedes the action—an infinitely protracted suspension perhaps not unlike the one in which the uproarious papers of Özkaya's *A Sudden Gust of Wind* buoyantly participate.[21]

21. Sigmund Freud, "The Moses of Michelangelo," in *The Standard Edition of the Complete Psychological Works of Sigmund* Freud, ed. and trans. James Strachey, Vol. 13 (London, Hogarth Press, 1953–74), 211–38.

DOUBLE (A THEORY)

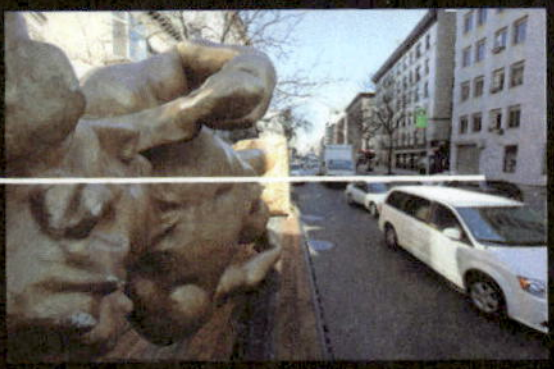

Christopher Eamon

1.

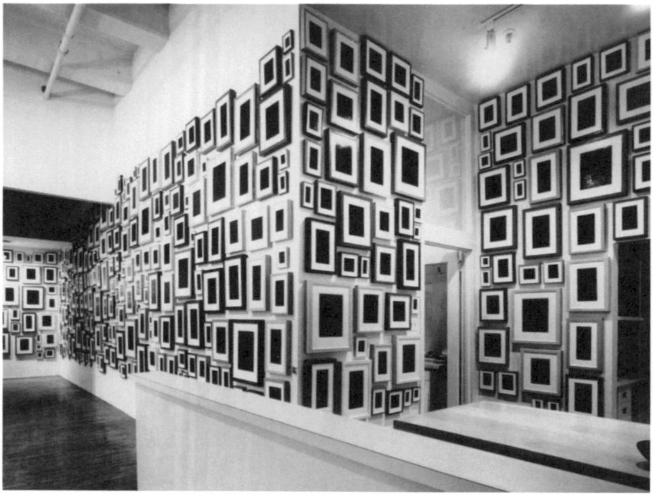

In the New York of the late 1970s and 80s, the concept of the double, or doubling, came to be used extensively in art and art literature, since then it explained the period's reflexive and extensive use of appropriation as a means of decoding and recoding the world of, in most cases, photographic images. The terms also become surrogates for "representation" themselves as works of art, language, and photographs all represent, and therefore double their referents. The newly translated texts of sixties French literary theory had shown representation to be inescapable and therefore highly suspect. The "double" is used in conjunction with and at other times in place of representation. Today's art language prefers the use of "figurative" to "representational" to avoid confusion with the now-loaded term, but captures only a tiny percentage of the latter's full meaning. Alan McCollum's *Surrogates* suggest both a massive culture of banality and a critique of commodity fetishism, as did Jeff Koons's direct appropriations of consumer products. The ambiguity between embrace and critique in the latter's work is, at times, played up, and we all know how that ended. The embrace won out.

2.

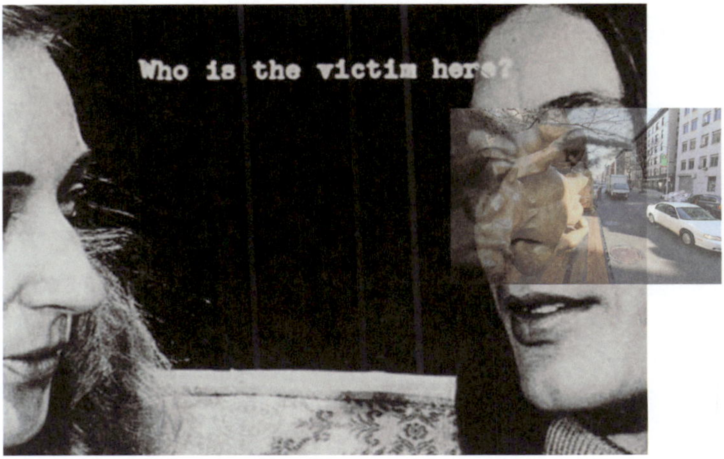

Writers and theorists, often categorized as those contributing to the Second Semiotics (DeLauretis, Mulvey, Kaplan, Silverman) in the late seventies, married semiotic insights in language to Lacanian psychoanalysis and Feminist theory. This powerful confluence retooled film theory as it was known at the time. In their work, the double as representation is shown to have a male, phallocentric bias as the camera in Classical narrative cinema doubles the point of view of a male gaze and is either voyeuristic or fetishistic. Avant-gardes are quick to respond in experimental filmmaking circles. Even pleasure – narrative pleasure – becomes suspect when derived from buried or hidden cinematic codes.

3.

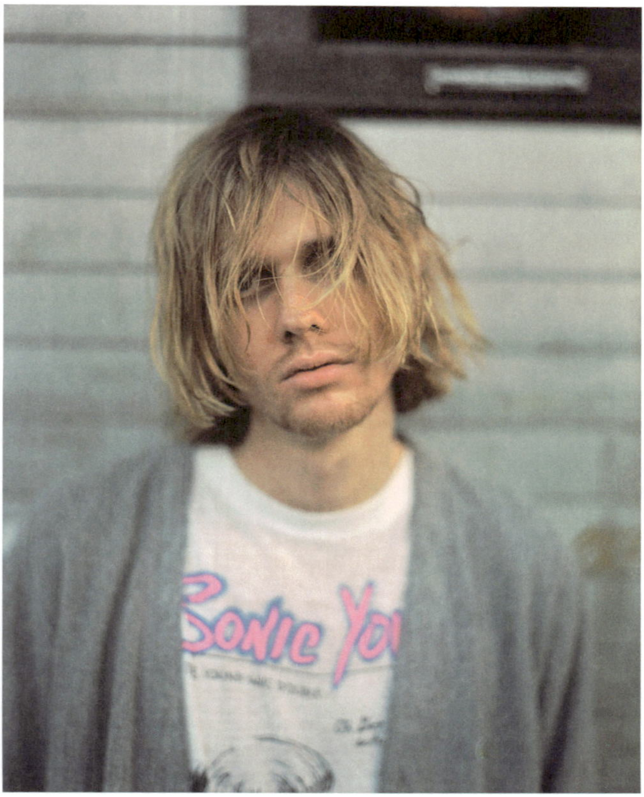

The storyline of E.T.A. Hoffmann's *The Sandman* goes as such: Nathanael, the protagonist, is warned as a child about the sandman. The eerie Coppelius comes to conduct experiments with Nathanael's father; Nathaniel sees Coppelius and associates him with the sandman. Nathanael is discovered as he observes Coppelius and his father conducting alchemical experiments; Coppelius wants to poke his eyes out, but the father saves him. Nathanael's father is later killed by an explosion during one of Coppelius's visits.

As a student, Nathanael meets the optician Coppola, from whom he buys a spyglass. Nathanael falls in love with the automaton Olimpia, whom he sees through this glass. Olimpia has been produced by Spalanzani (double of the father) and Coppola (the double of Coppelius). Nathanael observes a fight over the automaton and the eyes are pulled out of the robot's head. This fight between the two "fathers" repeats the fight that Coppelius and Nathanael's father had over Nathanael himself when he was discovered spying on their experiments. Nathanael goes insane and then recovers. When he is about to marry his fiancée Clara, they ascend the tower of the town hall and Nathanael sees Coppelius through his spyglass. He goes insane and tries to kill Clara. Her brother rescues her, but Nathanael jumps off the tower to his death.

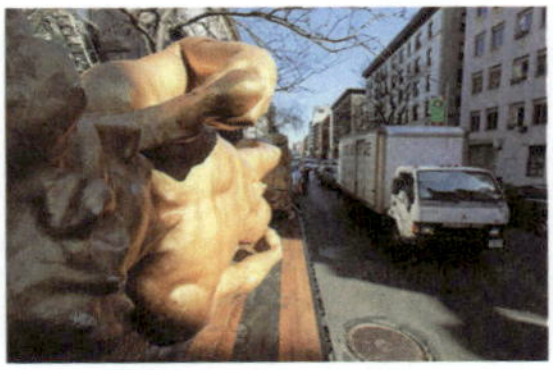

The spyglass is a kind of symbol for the revelation of a hidden secret – for a tool that makes the *heimlich* (secret) *unheimlich* (unsecret). It allows Nathanael to penetrate into the sphere of the "private," but also to gain access to the repressed parts of his own past.

4.

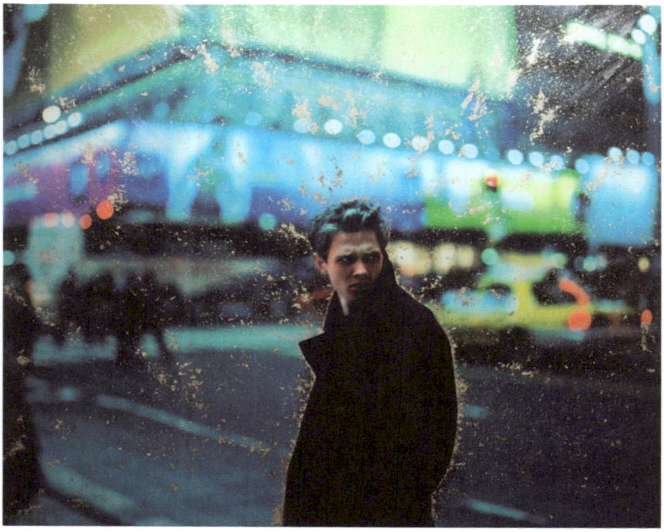

The buried or hidden is always at the root of double, for its meaning in some Romantic literature regularly features a doppelganger – someone who looks like someone else, or whose part in a narrative structurally mirrors or doubles a previous part of the same story, particularly one that involves trauma or terror. The automaton is also featured as an object with uncannily human characteristics. Perhaps the automaton can be seen as part-commodity / part-human, absent a subjectivity which through its absence or removal renders it uncanny.

Witnessing compels one to repeat until the normal returns. A hidden meaning may be discerned in our time in commodity fetishism, and yet traditionally, the appearance of the doppelganger represents a return of an evil twin, or a bad copy—a figure onto which one may project evil otherness, preserving the self as pure and unblemished, or allowing us to speak of one's own evil being.

5.

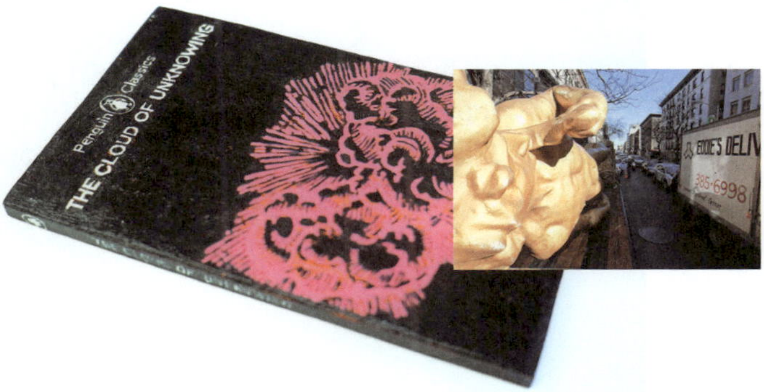

Speaking here and finally of the bad copy, one might see the aforementioned examples to be neither purely double, nor purely other. McCollum's framed pictures are only surrogates. They are roughly hewn from synthetic clay much like the hand-carved and painted books, objects, and furniture of Conrad Bakker. Here, the imperfections make the difference, which is as between the rephotographed

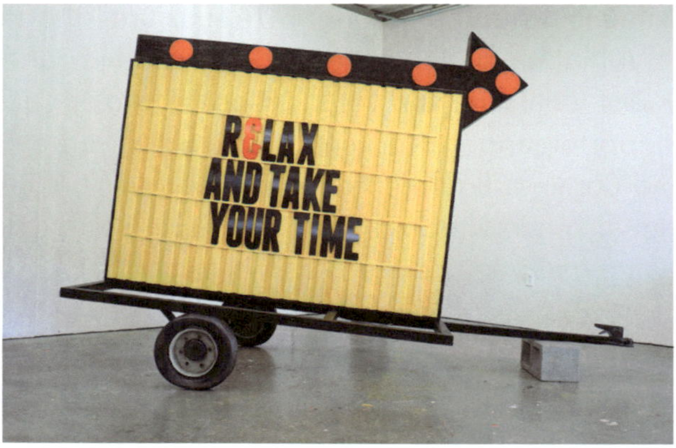

photographs of the early 1980s along with the appropriated object qua Duchamp, versus the hidden secret that begs to be discovered in the imperfect double. Psychosexual or phallocentric critique finds its abiding practice in the evil or imperfect twin.

MEMOS: ON DOUBLES, TRIPLES, AND SERIALITY

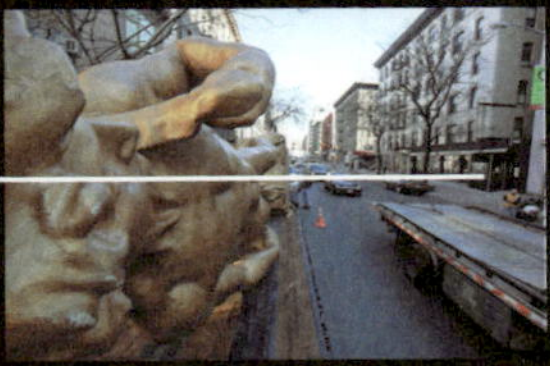

Alanna Heiss

Let us think about some doubles and duplicates. Warhol is important for so many reasons, because he is one of the only people I know in the art community who actually employed a double. Many people do this in the film industry, and I had the distinct honor of spending many evenings with people, never knowing they were doubles. Warhol had two people during the time that I knew him; neither of them looked very much like him, but they all talked like him, in this strange drone.

In music, we look at the absolute rules of Mr. Philip Glass, which are quite like architecture. We have inventions which were two voices and three voices. Anyone who has planned or studied those will know that it is very interesting to have two hands and two voices, and try it out with a third voice and a third hand (which means the foot).

Doubles are harder to find than you might think, but we do have the example of twins, and I would like to note that there are a number of twins in the art world. Perhaps there are many twins who are architects, but they only count as one person, thus being able to take twice as many clients. The twins I know for instance are the Starn brothers, who are identical twins. Then there are Os Gêmeos, from Brazil, who are identical twins as well. Then we have many fake twins such as the Chaplin brothers who are not twins but say they are twins, and people think they are twins, and treat them as twins, so I call them fake twins. Then we have the absolute-not-even-brothers-twins, which I will not discuss.

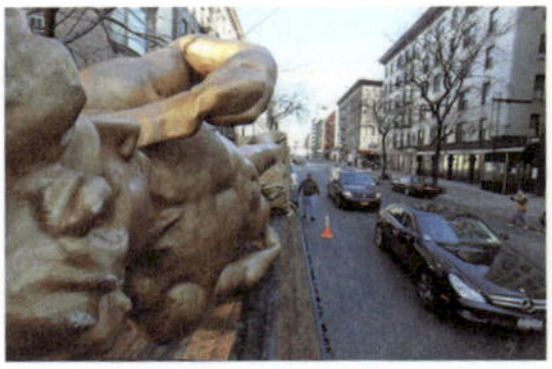

And now on to what all of us in the seventies love: serial repetition. Acconci built this very place [the Storefront] that we are in right now. There are the real serialists—the leaders, the gods, the gods of art, of serial heaven: Dan Flavin and Donald Judd.

In music, of course, we have Phil Glass, and his Music in Twelve Parts, a major serial piece written here in New York.

People in the rest of the United States had other serial music, but here in New York we refused to listen to it. In other domains, we have strange but exciting examples such as

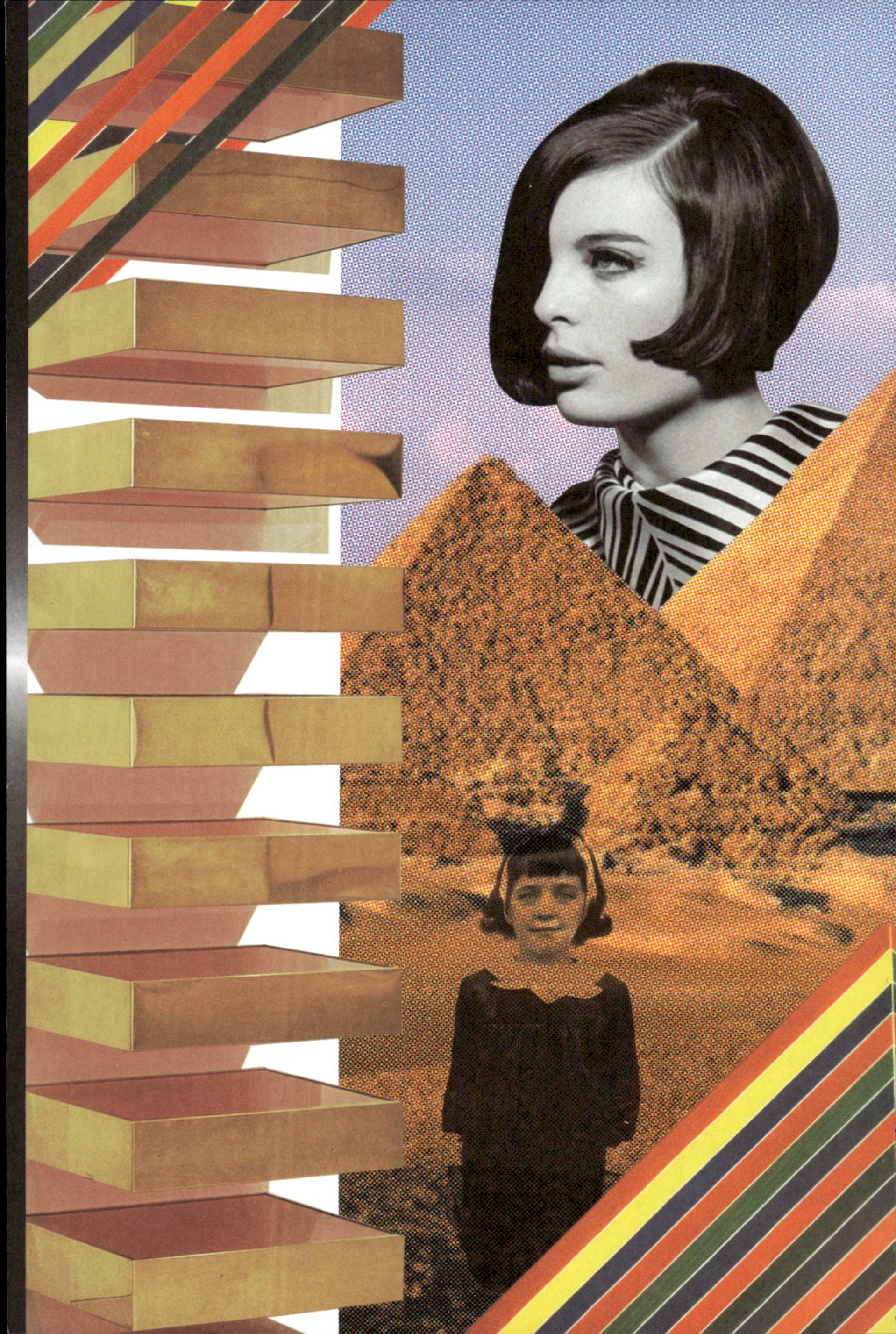

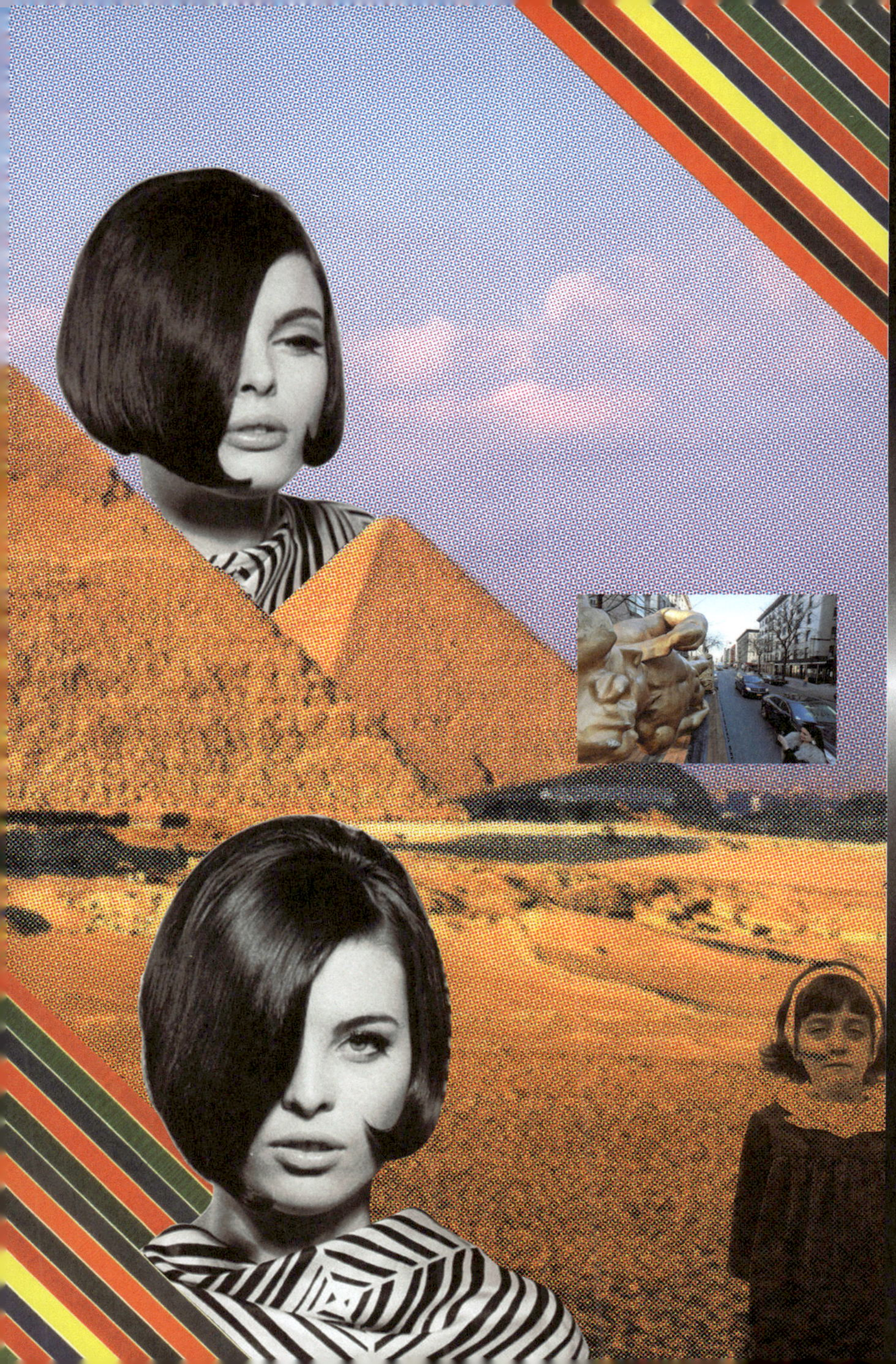

Elizabeth Taylor, a serial monogamist.

I close with the image of all of us as we will be in the future: cloned. As we will then all think the same things, these discussions on the copy will no longer have reason to occur. We will all be lambs before our god.

THIS IS NOT A MANIFESTO

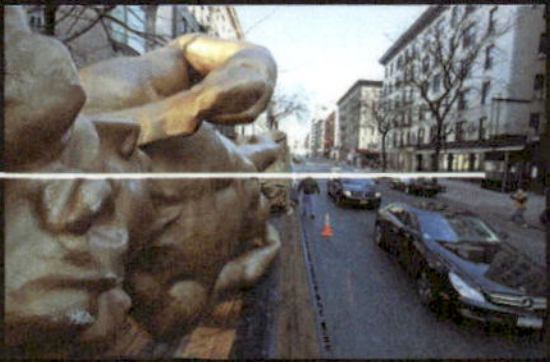

Serkan Özkaya

1. This is not a manifesto.

2. ON EXACTITUDE IN SCIENCE:
"In that Empire, the Art of Cartography attained such Perfection that the map of a single Province occupied the entirety of a City, and the map of the Empire, the entirety of a Province. In time, those Unconscionable Maps no longer satisfied, and the Cartographers Guilds struck a Map of the Empire whose size was that of the Empire, and which coincided point for point with it. The following Generations, who were not so fond of the Study of Cartography as their Forebears had been, saw that that vast Map was Useless, and not without some Pitilessness was it, that they delivered it up to the Inclemencies of Sun and Winters. In the Deserts of the West, still today, there are Tattered Ruins of that Map, inhabited by Animals and Beggars; in all the Land there is no other Relic of the Disciplines of Geography."[1]

3. The story above was the inspiration for Baudrillard's book, *Simulacra and Simulation*, which is the text that inspired the movie, *The Matrix*, which is the film that inspired the text, *Welcome to the Desert of the Real* by Žižek, and so on and so forth...

4. It's a beautiful idea to consider that the whole world or universe could be clothed in a simulacrum, and that despite all appearances, we are actually living on a map. We are all copies based on originals that are no longer around; the whole world has become a palimpsest. On the other hand, however, the

cartographic remnants here become relics of these lost originals—reminders of the fiction of unmediated, virgin territory.

5. If you were to cover the whole world with a map—a perfect map, whose size is that of the world—this very map would, in fact, need to be larger than the surface it covers. The map, the simulacrum, the replica, the double, the copy, would more or less need to be bigger than the original; with the degree of excess dependent solely on one's choice of sleeve material.

6. THREE DIMENSIONAL SCANNING AND MODELING:
My inspiration when I look into this arena is Dr. Marc Levoy from Stanford University, who developed his 3-D scanning and modeling system in the early 90s. His apparatus worked by sending laser beams from a source to the object being scanned—e.g. a pencil—measuring the distance between the source and the object from every angle, and creating a metastructure. This structure is then turned into a computer model of the three-dimensional object, in our case, of the pencil, from which a replica object can be printed. The print-out, then—the replica of the primary object—is the exact copy of the original; the exact copy of the pencil, except that it wouldn't write like our pencil because the material would be different. The print-out would be the sculpture of the object.

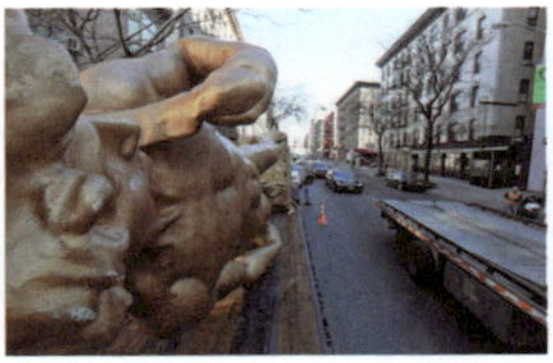

7. We manage the outside of the object, never the inside of the object. As Sartre wrote of Giacometti, the artist's method did not consist of shaping the clay constituting his figures, but rather, he shaped the room—the negative space of the clay.

8. SPEECH ACTS:
In linguistics, the "performative" exists in certain moments where an utterance takes on the character of an act. This comes from Austin's idea that most of our language is comprised not of propositions which can be evaluated as right or wrong, or true or false, or beautiful or ugly, but rather is in possession of an ability to do things with words. Under the right circumstances and with the right people, the formation of speech acts produces effects which can give shape to the world; e.g. "I now pronounce you husband and wife," voiced by the priest in the church with two people who are about to get married, or "I declare war," voiced by the prime minister to his Cabinet. Or, perhaps, a mother wishes to put a stop to her child's misbehaving. Not satisfied by the child's disinterested response to her initial plea, the mother might insist: "No, promise me," and until the child says "I promise," the act of promising is not fulfilled. Speech acts exist in a place where language intersects with the world per se. The speech act both authorizes the world and works within it.

1. Jorge Luis Borges, "On Exactitude in Science," in *Collected Fictions*, trans. Andrew Hurley (New York: Penguin Books, 1999), 325.

A CALL FOR THE REENACTMENT OF SOMETHING THAT NEVER ACTUALLY HAPPENED

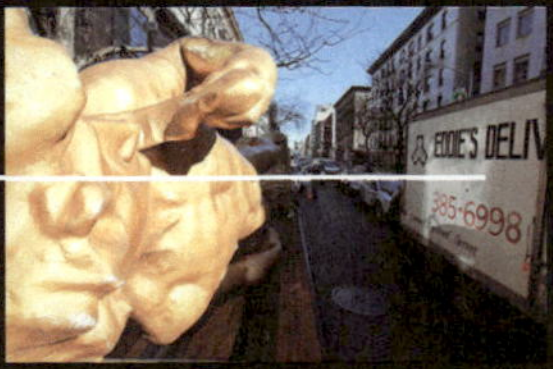

MARCH 06, 2012

Eva Franch i Gilabert, Ines Weizman, Cristina Goberna, Alanna Heiss, Spyros Papapetros, Hrag Vartanian, Christopher Eamon, and Serkan Özkaya

✦ I'm really interested in the evil twin or in the plan where something went wrong; also in the layers which make inanimate objects into objects of desire, or actually more than that: objects to be fallen in love with.[1]

✱ I'll just use the example of Daniel, the classic Freudian case of the evil twin and the Doppelganger. The double can be thought of as uncovering something that is not right. A late 1970s early 1980s example would be Allan McCollum's *Surrogates* or Yvon Rainer's films. When we go back to E.T.A. Hoffmann's *The Sandman*, the Freudian example of the Doppelganger in that story is manifest as a hidden evil. That way we can talk about other hidden evils in the world and the idea is inserted by semiotics following Lacanian psychoanalysis and the feminist theory. In the 1980s, however, this hidden evil was interpreted as revealing fetishism in the visual arts, whereas today, someone like Jeff Koons is not performing such a critique. The hidden doesn't necessarily have to imply a psychosexual element à la Freud which is based on the optical, on witnessing a trauma. Of course that's all there with Freud and you can see it in the optical and in the objects. On the other hand, I think the double doesn't necessarily have to deal with being about the optical or about sexual drives, but also of other ways of covering hidden meanings. I used the example of the imperfect copy because I wanted to get to the fact that Serkan's *David* is bigger

1
Refers to the love affair through spying glass in E.T.A. Hoffmann's *Der Sandmann*.

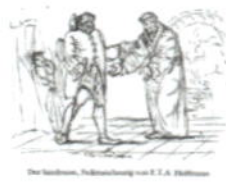

2
Referring to Christopher Eamon's Manifesto in this volume, pp. 71–78.

3
Designer Paul Cocksedge made an installation of flying paper sheets and titled it, *A Gust of Wind* a few years after Özkaya's *A Sudden Gust of Wind*. Cocksedge's installation was uncannily similar to that of Özkaya's and made no acknowledgment thereof.

4
See pp. 18–28

5.
Refers to Beagle2 where the Calibration Target Plate for the spectrometers was painted by Damien Hirst. In June 2003, all contact with the spacecraft was lost upon its separation from the Mars Express six days before its scheduled entry into the atmosphere.

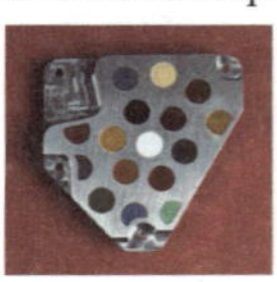

than the original and brings up other issues.[2]

⬢ What was interesting in Alanna's presentation is the idea of the double of the double, or the fake double, because if two in-affirmations create an affirmation they need to fake to create an original. I think it has to do with what you mentioned about how Andy Warhol had two doubles, but one double did not look like him at all. With two doubles, then, the fake of the double is somehow an original.

✪ I should correct that piece of information about which I'm not an expert. I actually didn't know that Warhol had two doubles at the same time. He was a serial doublist.

▲ Serkan, what does it mean to double the double which you had done with *David*? You had to recut the whole thing from a copy. Also, what did you feel about what I mentioned about the artist who copied your work without telling you and without asking permission?[3] Do you really feel that you can copyright a copy of a copy? I think you should.

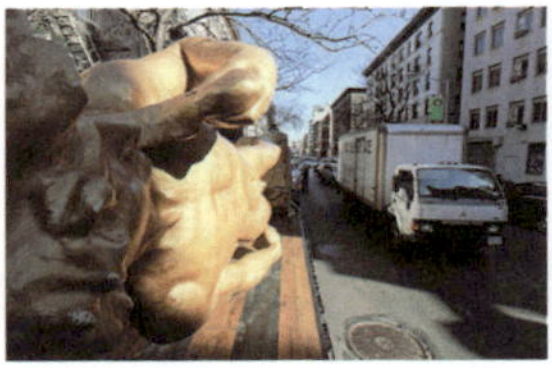

✦ I don't feel attached to *Double David*—as it is called now—nor to the second copy of it. But as Hrag said in his manifesto[4], even in the old days, the owner of the mold was the owner of the original. In this case, I think it is the computer model that is the mold; that is the original. The mold is the thing that is the closest to the idea. Several years ago, England sent a spaceship to Mars and it contained some of Damien Hirst's dot paintings. And it was a big debate about if this was the culture we were going to let leave to the outer world. And in the end when they sent the spaceship it crashed and went into pieces.[5]

If you ask me, this is not the way for objects or anything else to be left behind. It is the idea, or the thing which is closest to it—the computer model—which is a mold, that can never be destroyed. You can always destroy the outcome and then you can make another one.

And ideas will never have just one reincarnation; just like with this model of *David*, it will keep going and end up serial in some ways.

The use of repetition in Steve Reich would be a very interesting area to probe. It consists of certain imperfections so that it's not the repeating parts, but the intervals – the imperfect in-between spaces – that start to form a sentence.

Somebody showed the copy of *A Sudden Gust of Wind* to me[6] and I was really upset about it because that person did a horrible job. He produced the sheets from this really thick and heavy material.[7] And I think the reason behind that was because he wanted to sell them piece by piece, which didn't really make sense to me either. And then his whole installation was just a one day thing. I asked my lawyer about this and he told me: "You're not that kind of an artist who sues somebody who copies his work. You're an appropriation artist yourself!"

● I think there's an agency that still hasn't been copied enough. The claim towards objects of the past has yet to fulfill its entire political power. Therefore it's not a call for the copy, it's a call for the reenactment of something that never actually happened, yet it was there. And then when you go to the idea that you equate architecture to a subject, we animate those buildings and we make them the

6
Cocksedge's appropriation.

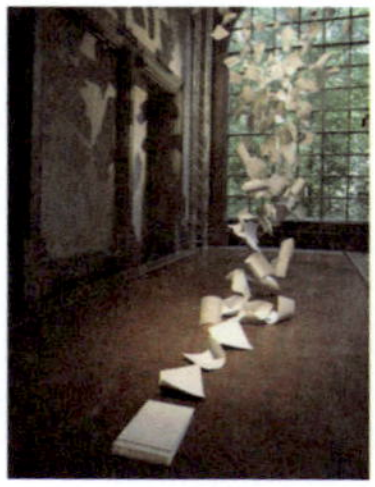

7
Corian.

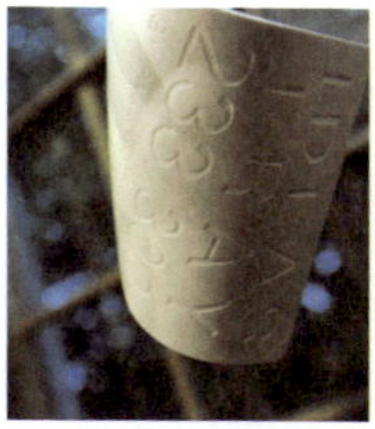

8
Özkaya's *David*.

9
Fake Industries Architectural Agonism's *Studio*, see second discussion, p. 119.

carriers of an agency that needs to be protected. It's not about the idea itself, it's about the performative qualities about what the things themselves are able to do. I'm trying to think about what we have here outside[8], because what we have is beyond the object, and beyond the sculpture and the destination; it's perhaps the entire image, the entire trip. And we've been talking a lot about what it is that this work contains. And it is a work that is on the move; it never stops.

We like to talk about copies and doubles and fakes, yet the constant thing we are trying to capture is that ineffable quality that the work contains—that which escapes the replication. And I'm trying to bring it back to architecture: What is it that we copy? I think that's the problem. In the past we just copied an image. And, in the case of *David*, it is such an obvious copy that you realize that it is not a copy at all. I think that is why your work is so provocative. Because by not being a copy it just unveils a different sort of potentiality. So my question is: How do we do it architecturally, if we can do it at all?

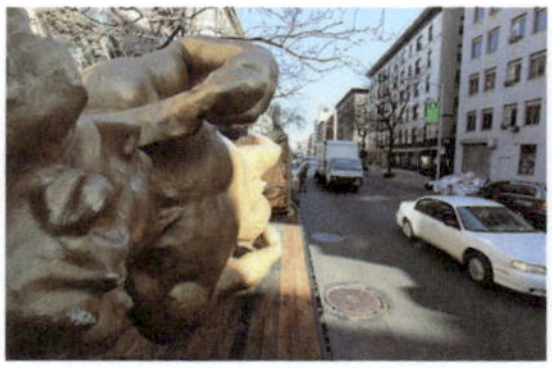

● Basically we don't do it. We talk about the replica; we only talk about architecture. The world destroyed was never there, only its potentialities. For example, we have just revealed the studio in China.[9] The Chinese government payed for that, so we included the architecture that was destroyed by the same government for political reasons. It had a really bad reputation in the architecture discipline. We believe in the replicas in the sense of their being an uncertain statement. It should be a copy, but also it should imply a political position, which means that the copy won't be exact. This lack of perfect correspondence will imply a certain statement.

✖ I have to thank Eva for inviting us to compose manifestos

because this is not what we normally write. The manifesto as a form of writing is obviously a moment when you project desire and imagine a world without limitations. As that was for me the activity in writing this, there are still hundreds of questions that come up in writing it. For me, it was an awkward thing to experiment with an unfinished Loos project – a project that was never realized. The interesting thing was that in 2008, seventy-five years after the architect's death, the copyright was free – Loos was one of the only of these early modernists where that was possible.[10] The work of many of the others is highly protected, particularly through foundations. The copyright expiration on Loos's work was for me a moment when questions of copyright in architecture turn into design questions. The architecture has to be protected from the architects in some way. What happens when there's the possibility that one can actually take those original drawings away?

⬢ I thought we were very interested in copyrights, but I found it really difficult to contrast with the issue of human rights. Copyright somehow implies property. Could it also imply merits? Also, copyright implies estate.

✖ Freedom of speech…

✦ Why do you think I should go after the copyright with my work?

▲ I didn't say you should, but I do think that in some way you should, at least in terms of the

10
See Ines Weizman's project with Loos's house for Josephine Baker, pp. 3–14.

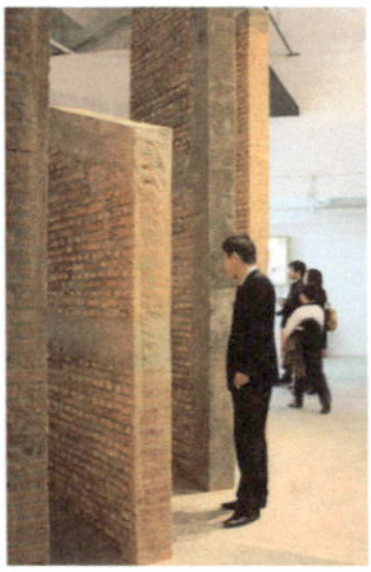

11
See p. 155.

12
Serkan had, in fact, seen and greeted his friend David Procuniar.

idea. That would be very tricky; very special.

⬟ I think a lot of people come to copyright from the wrong angle. The reason the copyright was invented was actually to reproduce an idea well and just. People were copying books and changing passages of books[11]. But, of course, we are talking about something hundreds of years in the past. So copyright wasn't just about money; it was about honoring the original intention and sometimes the idea. It was really interesting when you said we should use the right material. People don't think that's your work, because your work is nothing like that.

✖ This is only relevant if you really want it to equal you. If you said, "this is mine and I want to be well reproduced," then that kind of approach would have a completely different set of questions. For example, how does it look like? Why is this one better than the other?

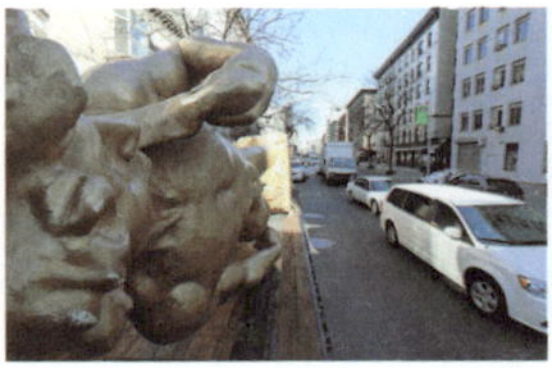

● There was a very specific moment tonight before we started the event. Serkan was outside and he was looking in the direction of *David* outside and he said: "David!" I had this really weird feeling that he was starting to talk to the sculpture. And I was a bit afraid and I thought maybe he was nervous, and he needed some advice from *David* or from the properties of it and he was trying to explain what he is about and what he actually wants.[12]

IN THE FUTURE THE DOME SHOULD BE A PLACE WHERE PEOPLE CAN FUCK FREELY

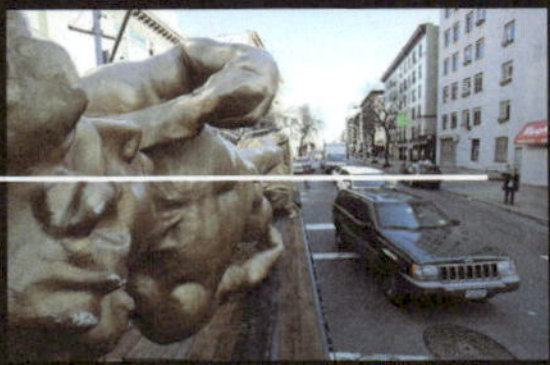

MARCH 28, 2012

Eva Franch i Gilabert, Liza Béar, Cristina Goberna, Pedro Gadanho, Urtzi Grau, and Serkan Özkaya

◆ You know the reason for my position; it's related to what's happening in architecture. The rendition of the models of masters is killing creativity.

✱ You mean the architects who have been copied a lot? They don't conceptualize the operation.

✚ So what were your objections?

✱ In architecture's postmodern moment, irony and quotation were the only operations allowed to happen. And then, we ended up with perfect originality. There were certain years where we were completely original – and that's so annoying. In Portugal, we call it the mangrove way of houses.

◆ It's something really widespread.

✱ Yes, they were changing the proportion…

◆ … and the model is Eduardo[1], because he is the master.

✱ No, it was because Siza[2] was too difficult to comment upon. First, Siza was already doing what I deemed as a very interesting appropriation. He was truly postmodern. He was quoting it all in a single work, making clear references to people like Adolf Loos, Alvar Aalto, Le Corbusier, and Frank Lloyd Wright. And there would be so much that this excess would become really beautiful. If you know a lot of architecture, you take more intellectual pleasure in Siza's work because you can recognize the references. If not, you would tend reduce his work to the experience of rather abstract spaces. Siza's work is so difficult to emulate or to copy. People were imitating the

1
Eduardo Souto de Moura, one of the references of the Porto School of Architecture, where he was appointed Professor. Souto de Moura was awarded the Pritzker Architecture Prize in 2011.

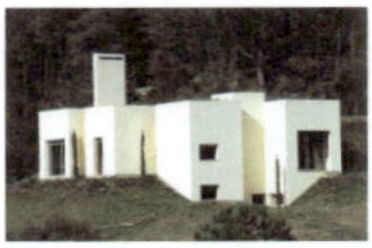

2
Explaining Siza's insistence on continuous experimentation in *Casabella* (July 1986), it was said, "Precisely for this reason his architecture can communicate to us an extraordinary sense of freedom and freshness; in it one clearly reads the unfolding of an authentic design adventure. In accepting the risk of such adventure, Alvaro Siza has even been able to bring to the surface, in his architecture, what one feared was in danger of extinction: the heroic spirit of modern architecture."

superficial aspects, which are the materials. They used white walls, marble, and oak, and that was it.

✦ Materials for "high" architecture.

✷ Yes, I'm talking about that. It is very basic architecture in terms of materials, but it's very rich in terms of spatial composition. But that composition is very difficult to imitate because it is so personal and so free. It requires translating the very free sketch into the built form – which is quite irrational; it's very difficult to translate. Looking at the sketches, you cannot understand why it is like this or like that. Why is this curve like this and not like that?

✦ A friend of mine wanted a large art work for his house and flipped through an art history book and found an early Jackson Pollock painting. (Img. 1) My friend then commissioned an amateur painter to make that big Jackson Pollock painting. So the amateur painter divided up the whole thing into squares and painted the whole thing one by one. It was a beautiful painting though, in a sense, it was not a proper imitation; it became something else. (Img. 2)

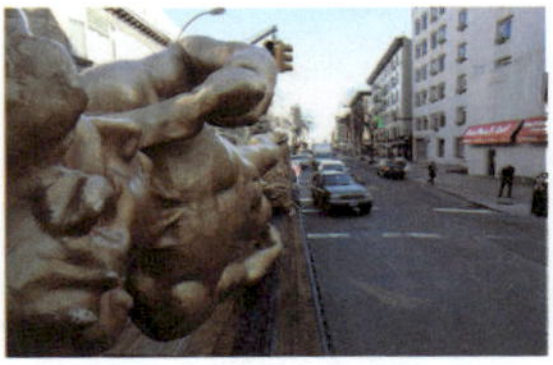

✷ You can of course emulate the conceptual production that is implicit in the work or you can pick up your own references and mix them. In this way, Siza uses the geometries and makes a combination from the different sources.

✚ The performative aspect of the Pollock painting would be hard to emulate. Everyone has a different anatomy. It would be hard to copy someone else's movements exactly.

✷ Of course, because this is a very intuitive work. How can you copy intuition?

◆ But the idea is to copy and do the same thing as what the other person did, or to copy and conceptualize copying rather than to copy someone. That's something different, isn't it? It's true, that you don't mimic a Jackson Pollock to become Jackson Pollock. But to make a Jackson Pollock exactly the same now is not to make a Jackson Pollock, it's to do something from the present.

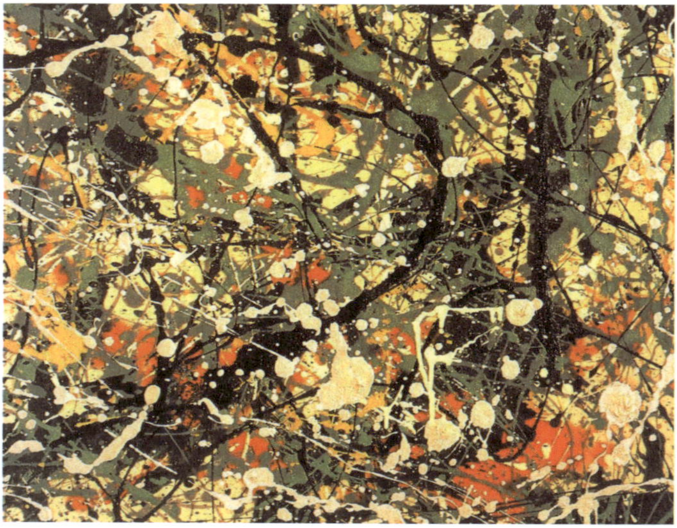

Img. 1

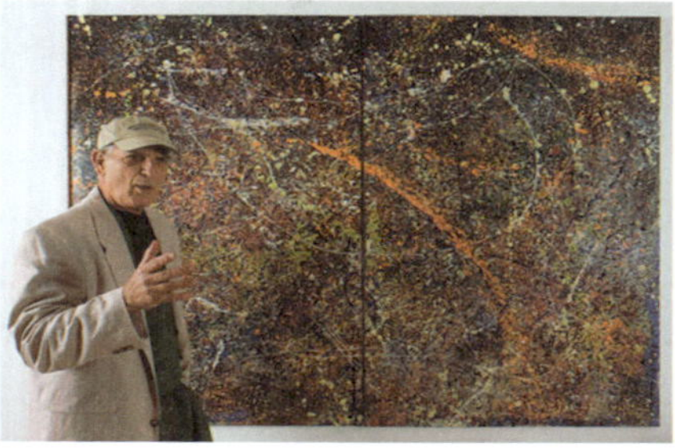

Img. 2

✸ But what if you apply the Jackson Pollock technique? That way you get something else that is…

◆ Then you don't get a Jackson Pollock; you get something else.

✦ My question is: Do you believe that architects who are copying a master do it with the aim to copy them or are they trying to make something of their own? What do you think the intention is?

✸ I think they're just emulators. Also, they are caught in this academic moment in which the thing becomes a model that everybody is supposed to follow because it's the best possible model. Rem Koolhaas becomes academic because people start to copy him. His work becomes an academic style.

◆ This is the story of the history of art and architecture. If you take any moment in the history of art you can see people who are important and people who are following them. Take cubism, for instance; it was a successful branch that became iconic. In any event, I don't think it's this kind of copying that we're interested in. I'm interested in copying as a conceptual operation, which enables thinking through the copying itself – like emulating something or borrowing something.

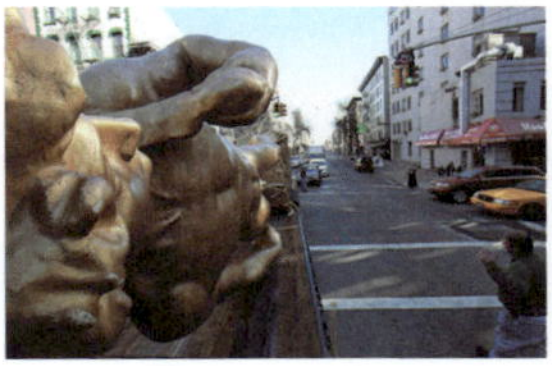

⬢ This is what's happening in Portugal and in the schools of architecture.

✸ Yes, and it is very poor in terms of its results.

✦ Is it like *cheating* in architecture?

✚ There's no satisfaction derived from it.

✸ I think not. I think people are trying to respond in a way that they think is fit for that condition. So they're repeating certain models. For a time, Siza was doing these houses that were very similar; together they were very Miesian. Following in Mies's footsteps, he didn't want to invent anything; he just wanted to find the perfect balance. Every house he did was a little more perfected than the previous one. He didn't want to make radical improvements. He just wanted to refine the thing more and more. He created this model of houses that was very easy to copy. In the hands of others, however, this process of refinement became an operation of rote copying and self-repetition.

◆ That model of copying has been there forever. That modern idea of copying has been circulating. This is a different one.

✦ That's what I was going to ask, because basically as soon as you have a reproduction, what is the point of copying in a creative system?

◆ Given the ease of reproduction, copying then becomes its own breed. That is why you and I acknowledge that an original is being questioned by the fact that it no longer has an aura. And the replica represents life – like trees on the surface of a canvas. They don't get printed on the canvas. They get the same treatment – the oil pads on it – so it's literally impossible to tell the difference[3].

✦ If we're not copying anything, we just pick up the readymade and displace it from its context.

● You used the same technique, didn't you?

✦ I did. This curator friend of mine, who loves this idea, thinks that once you can produce anything like that, you don't need the original anymore.

● So what kind of reactions do you get for *Double David*? I don't mean the first reactions,

3
Refers to the story from Pliny's *Naturalis Historia* (AD 77–79), where Zeuxis and Parrhasius staged a painting contest to determine the greater artist. When Zeuxis unveiled his painting of grapes, they appeared so real birds flew down to peck at them. But when Zeuxis asked Parrhasius to pull aside the curtain from his painting, the curtain itself turned out to be a painted illusion.

4
Refers to Goldsmith's *Soliloquy* (2001), where he transcribed every word he spoke for a week.

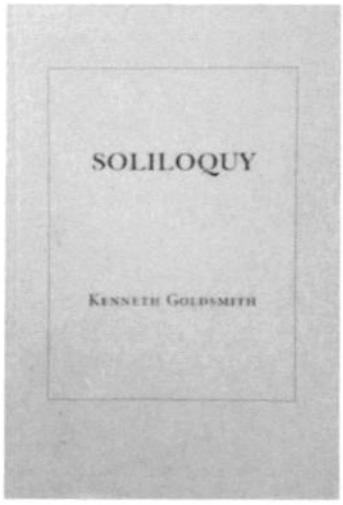

because I know everyone stopped their cars and took pictures and so on. I saw it in that article in *The New York Times*. **(Img. 3)**

✦ That was a funny article.

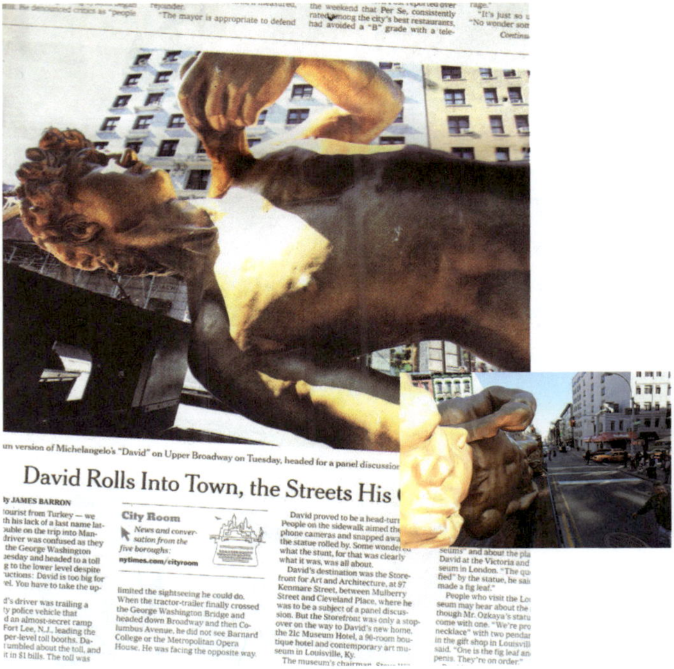

Img. 3

◆ Do you know this guy, Kenneth Goldsmith, and *Transcribed Stuff*[4], one of his most famous books? He gave this lecture in Princeton in front of all these professors, and I have never seen a more upset reaction by traditional literature people. They were really upset. There were people who were probably seventy-five years of age, standing up, saying: "This is outrageous, you cannot do that! This is absolutely outrageous!" And I realized that I had never imagined that such a project would actually have that kind of effect.

✦ What's the difference between Kenneth Goldsmith's efforts and mine? He rewrote *Ulysses* or *Don Quixote*.[5]

◆ You copied a Michelangelo.

✦ Yes. *Ulysses* is like an adventure. You start it and you're on your own for a month – it's a different kind of tale that you cannot imitate. But then again it is his effort that is like an artwork. Have you read his pieces for *The New York Times*?[6] His *New York Times* book? Have you read his *Ulysses*? Have you read his *Don Quixote*, or Cervantes's?

◆ No, but it doesn't matter.

✦ I have read Cervantes and James Joyce, but I haven't read Kenneth Goldsmith – because I understood his work when I learned about his process; that's his work.

◆ Do you have students of art going to draw your *David*? Are they going to spend two months drafting it? I don't think so. I would imagine, though, that around the *David* in Florence, you can see people there copying it for months.

✦ That one is not the original. You know that.[7]

◆ So, the original book is different because people spend more time on it?

● When we actually talk about your work as a copy, we see it as a strategy for producing something else. And this is perhaps where I see that *David* should remain as a traditional object; one that people are able to recognize immediately. And they say: "Oh, that's *David*." There is shock and awe that goes beyond

5 Goldsmith did rewrite certain pieces, but not high literature works. From his *Uncreativity as a Creative Practice*: "On Friday, September 1, 2000, I began retyping the day's *The New York Times*, word for word, letter for letter, from the upper left hand corner to the lower right hand corner, page by page.... The object of the project is to be as uncreative in the process as possible... I'm interested in a valueless practice. Nothing has less value than yesterday's news.... Retyping *The New York Times* is the most nutritionless act of literary appropriation I could conceive of. Had I instead, for example, retyped *Ulysses*, there would have been too much value, for *Ulysses*, as we all know, is a very valuable book."
In his book, *Fidget* (2000), Goldsmith attempted to record all of the movements his body made over the course of *Bloomsday*, June 16, 1997. He then transcribed the recorded narration he made of himself while performing the task.

6 Ibid.

7 *David* statue on display outside of Galleria dell'Accademia.

the recognition of actually seeing *David*. It's like when someone keeps knocking off preconceived ideas in order to produce some other work. The copy is a means, not an end of your work. And I think that's different from what we are talking about here: taking a copy as an end product. I think that has to do with how you guys refer to copying. What are the performative qualities of the copy? How many different taxonomies of copying can we find? Copy as a "versus," copy as a method, copy as a destination, and as *David in New York*. You have produced some copies previously, where the process was different and the work would become something else, such as in copying the cover of *The New York Times*[8], where the work is obviously copy-as-a-process, but it is clear also in the object that it is a distinct work. There is always a quest for clear differentiation. To only use the verb "copy" in your work is actually one of the ways in which to *not* talk about the work.

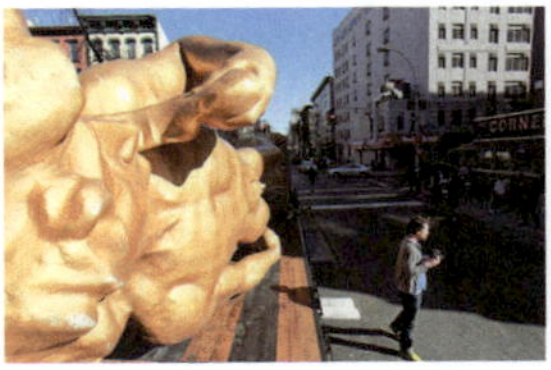

✦ The *New York Times project* also considered the newspaper as an object. It was a copy. It was a newspaper, but it was also the news for that day. So what Kenneth Goldsmith does is precisely what Pierre Menard did. Pierre Menard's story...[9]

✸ I don't know if it's by Borges, or if Borges is referring to it, or somebody was referring to it while talking about Borges – the story that if you had one thousand monkeys just writing

8
Refers to Özkaya's *Today Could Be a Day of Historical Importance*, where the artist hand-rendered the cover page of the newspaper, and it was used as the actual page in the printed newspaper the next day. Done with five newspapers from 2003 on, including *The New York Times* in 2009.

9
Pierre Menard, Author of the Quixote by Borges is a story about the author who rewrote *Don Quixote* "word for word and line for line. ... He did not want to compose another *Quixote*—which is easy—but *the Quixote itself*. ... he never contemplated a mechanical transcription of the original; he did not propose to copy it."

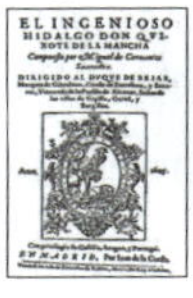

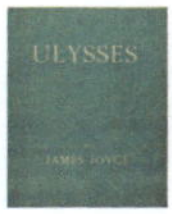

automatically they would eventually write all of the classics.[10] You'd have all the master works. That's why I think it's important if Kenneth Goldsmith was reading his text in straight lines or...

◆ Actually, he reads it vertically, but he goes page by page. So the first page of the book starts with a mini text with the title, *The New York Times*.[11]

✦ That's fascinating. Imagine him reading it.

◆ It's absolutely amazing.

✦ Fascinating, but reading it for yourself is brutal. He wakes up and searches for his pulse for the whole day.[12] Every single movement is recorded. It's a great...

◆ Do you think you can read it by itself? Just like that?

✦ Of course!

◆ Poetry. If you believe in something like that, perhaps it makes more sense?

✦ Yeah, absolutely, are you kidding?

◆ Those pieces are to be performed, and at the same time, they are conceptual operations. Like the poetry of Picasso...[13] There's poetry like that. You don't have to read it; you need to consume it. There's so much stuff you have to deal with in terms of the material as well as the meaning.

✦ You have to have balance. I understand you propose aspects of performance in a certain context where the concept is clear as well, but there's a fine balance between James Joyce's oeuvre and Kenneth Goldsmith's.

10
To the idiom, "that a half-dozen monkeys provided with typewriters would, in a few eternities, produce all the books in the British Museum," Borges added: "Strictly speaking, one immortal monkey would suffice," and then imagined the contents of the *Total Library* that this enterprise would produce if carried to its fullest extreme. He also wrote *The Library of Babel* in 1941, which describes an unimaginably vast library consisting of interlocking hexagonal chambers, together containing every possible volume that could be composed from the letters of the alphabet and some punctuation characters.

11
Refers to Goldsmith, *Day, The Figures* (Berkeley: Small Press Distribution, 2003).

12
Fidget. See footnote 5.

◆ Yes.

✦ You say it's beautiful, but then it works in a different way than you would expect from a plastic piece.

◆ The images of the Empire State Building[14], for example, are important because the film has no plot, but it's not consumable in its totality because it's too long. Are you going to watch the movie for eight hours? That doesn't destroy the content of work.

✦ In the particular case of *David's* larger-than-life replica, it depends on the context. It depends on the architectural and social structure in which it's placed. Have you ever seen a picture of it in a museum?

◆ No.

✦ It's always outside; it's always in a different context, which is part of the idea as a whole – it's performing itself. Personally, I wouldn't look at it as a wonderful copy of Michelangelo's piece and try to inhale it like one would with the classical sculpture. I wouldn't do that also because I'm not so keen on the originals.

⬣ Especially in the *David* case, the object is so striking and big that you cannot deny its presence. Be it a copy or not, it doesn't really matter, at least in that specific case. Because this *David* is huge it stops traffic in the city.

● That's true when you first see it, and then you get over the scale and recognize it as an historical image, and then you get really into the actual object. What's extremely interesting is

13
Gertrude Stein wrote in her book *Picasso* (1938), that, "The egotism of a painter is an entirely different egotism than the egotism of a writer. ... This was his life for two years, of course he who could write, write so well with drawings and with colours, knew very well that to write with words was, for him, not to write at all."

14
Refers to Andy Warhol's *Empire* (1964), a silent, black-and-white film. It consists of eight hours and five minutes of continuous slow motion footage of the Empire State Building. Abridged showings of the film were never allowed.

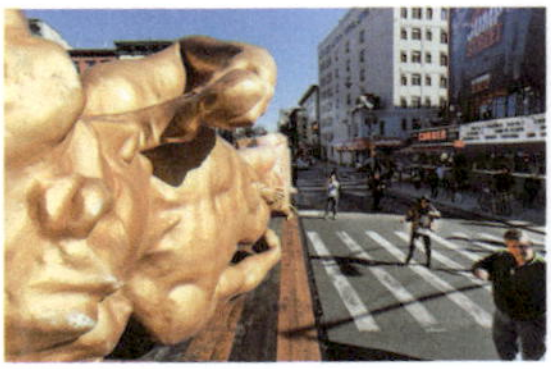

that it goes from concrete to abstraction, and then to some other stasis. It feels like saying: "Oh, it's big! Oh, it's art!"

● It has certain expectations to stand up to. I was just there for a while and two people stopped and...

● But let me ask you guys...

✚ Don't forget that he's holding a stone and a sling and he's pointing it towards Rome.

● That's interesting, because actually the sexuality in the piece made me see other things in that stone.

✚ I'm not talking about the stone; I'm talking about a person who was chosen to stand against Rome – that's why he is standing with the sling and the stone. If you ignore those aspects of it, then you're ignoring the very reason why Michelangelo was commissioned to do that piece.

● But you don't think those aspects are important in the context of the figure itself even if he would have changed a small aspect. This is why I say the copy is an idiom.

✚ I thought it was relevant in today's context – to notice that he's carrying a stone.

● I'm trying to understand, because you have been obsessed by repetition for a long time.

◆ We have three different things that we do in a week: we teach, we research, and we do projects in the office, and a lot of them intersect in operations. Modern copies; we mainly call them replicas.[15] At Columbia, we were able to find fivety-five different ways of

15 Fake Industries Architectural Agonism. See their manifesto, pp. 15–17 and the second discussion, "Let's ask the question differently..." pp. 123–58.

16 An adaptation of the book *The Hobbit* dating from 1953.

17 Nicolas Bourriaud, after his emblematic book *Relational Aesthetics* (see page 129 in the following discussion), his second book *Postproduction: Culture as Screenplay: How Art Reprograms the World* (New York: Lukas & Sternberg, 2002) analyzes new ways of art-making and makes concrete analogies between DJs and artists.

copying, or replicating, through which our students are working in different ways. These methods ranged from the musical *The Hobbit*[16], to the mash-up technique, and to post-production work.[17] In our practice we make stuff, we appropriate, we copy, and we reproduce. In our specific pieces and projects, we work in different directions.

● You take an ideological position in which you are going to copy certain things.

◆ In Borges's story, Pierre Menard copies a story word by word, and that is what is so different. It would be a painful experience to transcribe *Ulysses* and write it word by word.

● But that is a performative experience. It's a book that stays in a certain way as a reference; in other words it is the ghost of a performance.

◆ No, what are you talking about? No one ever read the entire book. It never happened. Who is going to read 600 pages? It is not a performance, it is a book.

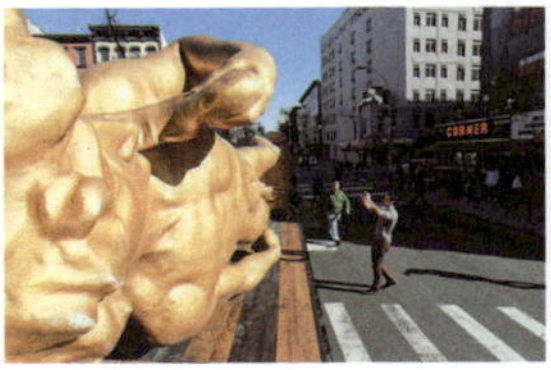

● Let's try to articulate...

✷ But it's funny, Goldsmith says that in fact, all books are performances. They are the end result of a performance, which is the act of writing the book.

● That is not a performance, it's a process of going through a mental state.

✦ When you read *Ulysses*, that book is amazing. When you read this other *Ulysses* by Goldsmith, you try to read it in the context in which he actually decided to copy the book. That is the difference that he puts between literary and established books.

◆ As if the first *Ulysses* has no context and the second has a lot of context? But that's ridiculous!

⬢ What about imagination?

◆ There are no separations in art that are assumed to have value.

✦ A picture is yet another thing. It is another picture that is a work of art.

● Imagine that you reread a work through the eyes of somebody else; it's very powerful as a literary exercise. I don't think the work being read loses its power in that case.

✦ But I do believe in the difference between the literary effort and painting the picture of something, because a picture is not the same as a book.[18]

◆ Painting the picture is a kind of process that is completely removed; but copying a book is a physical process.

✦ But in that process of copying, when one aims to copy, what assumes value are the spaces of separation – the differences between what the original is and what the copy is; it's not about doing something. There are many things that are removed from their condition as the original in order to still produce a replica that separates itself on so many degrees. You can take those degrees as generative moments, and they are full of content. Meaning that shifts does not have meaning. If all the meaning shifts, nothing will have meaning. The meaning itself has to be read outside. The copy is convoluted.

◆ When one copies a text, one is changing the typeface, the font, etc. Poetry and literature have to do with typesets and the way you compose pages.

✦ You think that reading the first edition of a literary work makes more sense than reading

18
Milan Kundera talks about this almost Kantian difference between the works of an artist and the works of a writer in his collection of essays, *Encounter*: "...I say 'I love Joseph Conrad.' And my friend says, 'Me, not so much.' But are we talking about the same writer? ... Is that the situation in all the arts? Not entirely. If I told you that Matisse was a second-rate painter, it would take you no more than fifteen minutes in a museum to see that I'm a fool." Trans. Linda Asher (New York: Harper Collins. 2010), 57–58.

19
According to Borges's short story, Menard's research and labor for *Don Quixote* is not among his nineteen pieces that survived: "...'My intent is no more than astonishing,' he wrote me the 30th of September, 1934, from Bayonne. 'The final term in a theological or metaphysical demonstration — the objective world, God, causality, the forms of the universe — is no less previous and common than my famed novel. The only difference is that the philosophers publish the intermediary stages of their labor in pleasant volumes and I have

another copy of it?

◆ Yes, it's a material transformation – of cover size and paper stock. It's about a canon of beauty and you see it as pictures. It matters.

✚ This starts to ignore why the sculpture was commissioned and how it was installed and why it faces in a particular direction.

✦ This particular object was chosen because it is the signifier of sculpture in our civilization. It has nothing to do with *David*; it has nothing to do with Michelangelo; it has nothing to do with Rome!

◆ Can you say the same about someone transcribing *Ulysses* because it's an icon for twentieth-century culture?

✦ It changes so much from the initial work. In the end, *David* is just an object – just a shell. *Ulysses* is an adventure you go through. Even if you have the shell of the book, you don't take away everything from *Ulysses*.

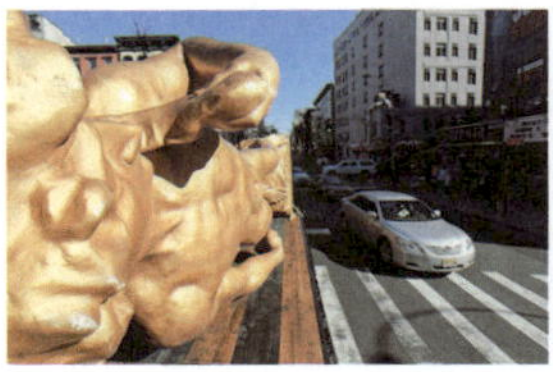

◆ Are you saying David is just a shell?

✦ It is a shell. The work of art is what it looks like. There is no inside to it.

✸ I think the beauty in Borges's story is in the fact that it leaves much to the imagination. Imagine that this guy was fascinated by *Quixote* and being two hundred years ahead, he actually managed to copy, no, to rewrite, the story again with the same exact words – you have to really imagine it.

◆ And find a kind of mental state that makes sense of all the words.

✸ But this is obviously a phantom. This is the beauty of the story – that it's an actual fantasy!

◆ He was able to write two and a half chapters.[19] Menard first attempted *Quixote* by trying to learn the Spanish of a Christian crusader of the seventeenth century; to conceive of being in a fight against the Turks – to become Miguel de

Cervantes.[20] Was this serious? It's an anachronism. The problem – the difficult operation – is to write it from the present; to make every word make sense, while nonetheless garnering the validation of the past. And therefore, when copied, the body of those words increase information because each of these words absorbs all the past – all the history that has happened – including the writing of *Quixote*.[21]

- I believe this is akin to the political agency of certain buildings or architectures that have not been able to fulfill or realize their potentiality, and therefore should be copied – copied in a contemporary context to see if that work still makes sense now; right now. To rewrite something in the same exact way in which it was conceived would mean that the context had not changed – that the context which brought the work into existence is, in a certain condition, the same.
- But why in the same condition?
- There is always change, but if we suggest that a work can retain validity across time, then clearly it responds to some contextual element that has, over the years, remained the same. We copy those objects which maintain their value across a long duration precisely because their charge is still valid. Or, let's say, a work is a reenactment in itself, and therefore, it still possesses the potential to produce the kind of change it once produced in its initial setting. When you were presenting your manifesto[22],

resolved to do away with those stages.' In truth, not one worksheet remains to bear witness to his years of effort. And the final work or rather 'a kind of palimpsest' of the very work is given to the storyteller in order to be reconstructed, 'unfortunately only a second Pierre Menard, inverting the other's work, would be able to exhume and revive those lost Troys...'"

20
"The first method he conceived was relatively simple. Know Spanish well, recover the Catholic faith, fight against the Moors or the Turk, forget the history of Europe between the years 1602 and 1918, be Miguel de Cervantes. Pierre Menard studied this procedure (I know he attained a fairly accurate command of seventeenth-century Spanish) but discarded it as too easy." Ibid.

you were saying that there are some works which have a certain political power – works that still have validity in our age. So, while of course many things have changed, perhaps in some situations the political orders haven't changed at all.

● Sometimes what happens is that a thing could have a value, but that value is only produced through a reenactment. This might be the case with things that have been forgotten or destroyed for whatever reason.

● There's an exhibition coming up at Storefront where we are just taking 101 projects for New York that never materialized, and we're going to copy them and project them into the present time. (Img. 4)

✚ So you are going to complete them?

● Not necessarily. We are going to see what happens.

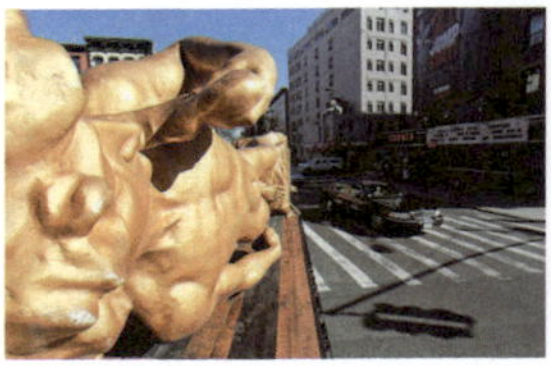

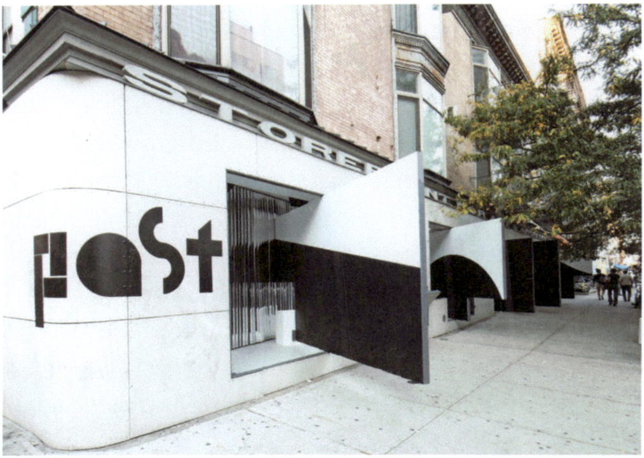

Img. 4

21 "Cervantes' text and Menard's are verbally identical, but the second is almost infinitely richer. (More ambiguous, his detractors will say, but ambiguity is richness.)" Ibid.

22 See Fake Industries Architectural Agonism's manifesto pp. 15–17

✚ Are you going to copy the proposals?

● We are going to try to understand why they didn't happen – what was it that they were proposing that was going to be able to change the context and the circumstances – and try to understand what would be the validity of those ideas in our contemporary context. And then, as a third stage – as a test or validation of that process – we will take these past projects and bring them into the present, and then project them into a future.

✚ How so?

● How so? With the help of those ideas, we try to create a context! The idea is always about the context, right?

✚ Wait, you are going a little fast. How are you going to project them into the future?

● You need to create a fictitious context into which these ideas are going to be able to validate themselves, therefore...

✦◆ Like what?! Give an example.

● Take, for instance, Buckminster Fuller's *Dome* over Manhattan. (Img. 5) This is a project that existed within a certain context, and is bound with certain desires. You could bring this project into the present by giving it to architects X, Y, and Z, and saying, "You know, these days we are worried about ecological conditions and the idea of sustainability or recycling, and now we are also considering CO_2 output, etc." Here, then, is an entire new set of words, concerns, and constraints alongside the ideas of property or collective parts that will come into play.

The stage where we bring this project into the future is one in which we can imagine the problems and constraints that we might have in the future. This could mean, for instance, that in the future the dome should be a place where people can fuck freely, or it could mean the context that one ascribes or in a certain way gives to the future is phantasmagoric or utopian or...

◆ There is a real organization that we use in our work. We

borrowed an event from a doctor in the seventies, from a moment when there was an excess of medical research papers. Right now, there is so much production of research in medicine, that there is no way for people to actually observe and evaluate what is going on. That means we have too many papers, and not enough time for articulating any new piece of knowledge. The scientific method requires presence and hypothesis, and then you test the hypothesis. And then you take the information on what's being produced and recombine it and put it in a certain context so that one may discover stuff. With this certain scientific method this doctor got about twenty discoveries; they were completely original. This made him realize that such discoveries can only occur in moments of excessive production; moments when you don't have the time to observe the knowledge that is being made public.

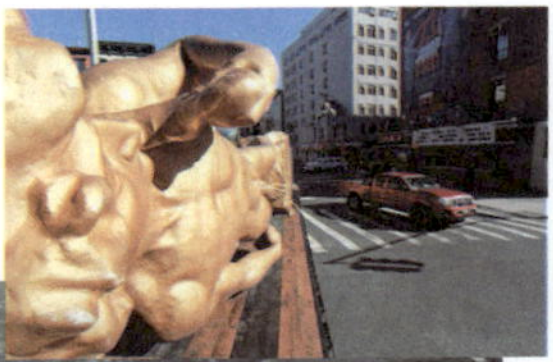

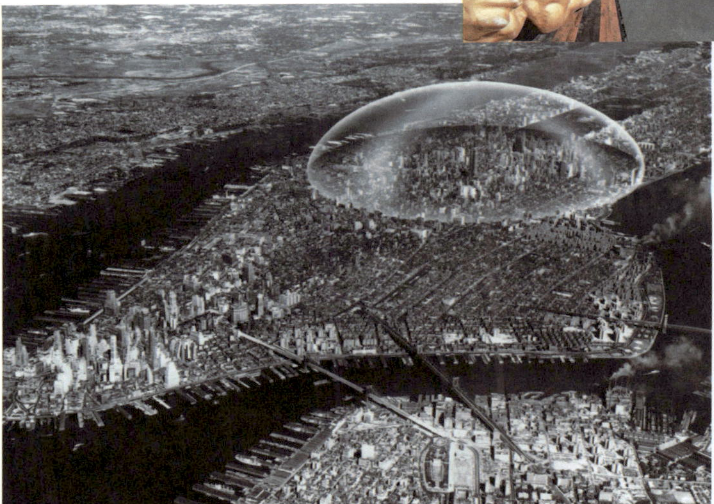

Img. 5

✸ I would say that is what happened with architecture over the last fifteen years, and we now have more resources than we can explore. Today, I was doing my exercise which I do every day; that is, going through all the newsletters that I receive – I assemble them, and I am amazed every time. Currently I'm impressed by the quality of innovation; by the quality and pure capacity of what architects produce. Every day there are at least five to ten new projects that are of value.

● According to which principles? That's the problem. We don't know...

◆ We are looking at the most commercial ones.

✸ I'm just giving this example, because there we lack the capacity to recognize and process all of this work.

It doesn't mean that there are five or ten good ones; there still are maybe two hundred thousand projects that are very bad. Compared to previous moments in history, when architecture was recognized through the same qualities in certain places – qualities of innovation, qualities of progress, and so on – it seems that today very little of what we see can be evaluated in such a manner.

✚ I was thinking that as well. I would like to ask you whether these good, valuable projects included projects for educational purposes?

✸ There are a lot of projects for schools.

⬣ I will have to disagree with you. What you're saying applies not only to today's architecture, but also to the bulk of our new cultural

23 Wikipedia claims a similar definition: "An architectural structure is a free-standing, immobile outdoor constructed element. The structure may be temporary or permanent."

production, from blogs to uploaded videos shot by a smart phone, etc. There is no criteria for comparison at all, though I'm sure that we are close to much more architecture right now than ever before.

● There is an amazing body of work out there, but if one actually has thoughts of what architecture should be, or what architecture could be, I don't think we are actually in an era of endless production at all. Everything that is speculative in nature is so banal. In terms of political and social aspirations, we are so lost.

The moment that you are actually able to call something architecture you are already giving it a judgment of acceptance—you automatically think and define the realm of what architecture actually is. You don't consider architecture as pantomime or accept just any building out there.

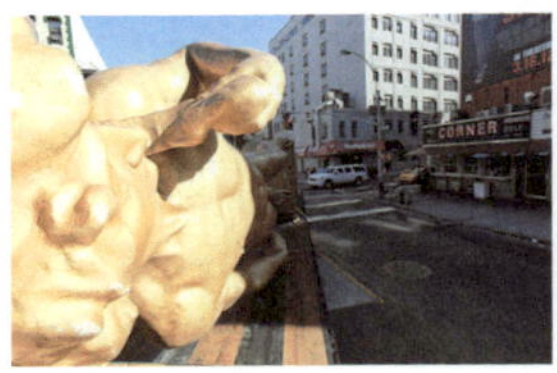

◆ My criteria to understand architecture may be based on that.

● This is really basic; your aspiration forwards the production of architecture, and so I don't believe you. I know that you have different aspirations, and they're not simply about faking something.

✚ The NYPD definition of an architectural structure is sides and a top.[23] Anything that has sides and a top is considered an architectural structure.

● A building may be a structure, but not necessarily an architectural structure.

◆ I'm far more interested in the NYPD's definition of architecture than those put forward by architects.

● I think you are being too mild and too kind in accepting. Is your work that of an ethnographer, where you try to bring in history and try to find the agencies of architecture in the time when it was made or try to acknowledge its potential—a business of unearthing the latent potential that a work carries, unbeknownst, perhaps, even to its architect? Is that a project that you're interested in doing, or is it rather that you believe that the works that are being produced now are pretty amazing and so you're just going to reenact them?

◆ No.

● Because you are both historians.

◆ As you know, Cristina and I only disagree on certain details, but the basics of Fake Industries are our agreed practice. My private interest in fakes is that we are able to produce a fake that doesn't need an original to be copied.

⬢ I disagree!

● This is fascinating! Tell me what it actually means, please.

◆ In the discussion about originality, the architect is not part of the fake, yet is able to copy. It's accepted as an oxymoron because it is assumed that questions of originality are innate.

● Give me the new terms in which we can discuss the idea of the fake, authorship, and originality. Give me the new terms!

◆ A fake: a copy that doesn't require an original to validate itself!

✦ I used to look for the particular object or set of objects in which every one of them would be an original. One of the answers I came up with is the newspaper project.[24] On the other hand, we could also consider a copy of

24 See footnote 8.

25 Fake Industries Architectural Agonism proposed the rebuilding of the artist Ai Weiwei's studio's facade—infamously built and immediately demolished by the Chinese government—a collection of *Domestic Dreams* in the form of four architectural designs—thirty models—displayed on a table, and a film that portrays the orgy of construction and destruction that took place in the time between *La Strada Novissima* in Venice 1980 and *The Street* exhibition in Shenzhen 2011.

a book of literature. There is no apparent difference between one copy of *Ulysses* and another one from the same print run, and each one of them is an original that provides you with the original reading experience, or at least the potential thereof.

● Could you show me images of the exhibition in Shanghai – of the copy?

◆ "There is not original work in this installation."[25] [sic] To make that comment in the context of art would be difficult.

● I don't think so.

◆ In architecture, you can't go around saying, "I don't produce original buildings."

● Why do you say that and then not do it?

✦ It's the gesture!

◆ You copy a wall, you copy a plan. We get these commissions that are connected to preservation. We are in a neighborhood and the rules of preservation say that we have to keep the walls as a certain type. I am interested in people who spend their lives updating works of architecture.

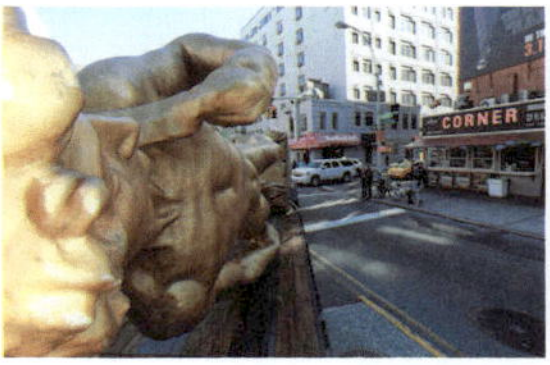

● So what is it that you are copying?

◆ What do you mean?

● An author who wants to translate a text tries to enter into the conceptual context of the original piece, and tries to break it sometimes, and keeps making introductions where he actually excuses himself for the liberty he takes in order to make the translation. So, I'm not asking about the actual pieces; I'm asking about the introduction towards the leaps, that you are taking from the original works in order to produce your work. I want to know about the leaps; what are the leaps that you are taking?

◆ Well I mean it's like…

- … in the works. The idea of typology has problematic social implications.
- Let's use the example of the translator. Let's say a guy wants to translate *Ulysses* from English to Spanish; he has to decide which letter in Spanish to use. A translation is not just about having the same amount of letter 'a's in that work; quantity would be ridiculous as the criteria for a good translation.[26] The translation is an appropriation that you undertake because you recognize the value of the thing that you are translating.
- In a good work – the Shanghai project that we are talking about, for instance – what leaps and what literal moves have you taken and what modes of separation do you produce in order to bring it into a different space of translation?
- Are you talking about the facade?
- Yes.
- In the case of the facade, for example, one of the things we want to say is, "We are not interested in producing another sensitive beautiful facade, like a super designed structure or another original work." Your second question was, "Tell me what you do," right? In our reply to the Chinese government, the idea was to recover some material they had destroyed. We found some pieces and leftovers of the original wall and constructed some artifacts out of them. On the one hand – for me, at least – the endeavor was to take a position on destruction and construction, which in this

26
In his 300-plus-page novel *La Disparition*, author Georges Perec does not use the letter "e". Neither the English translation, by Gilbert Adair, nor the Turkish translation, by Cemal Yardimci, uses the letter "e," except in the author's and translators' names. In all these languages the letter "e" occurs extensively.

case, has to do with the real estate bubble; from Spain to here, and from here to China. Sometimes production is mixed with the politics, and in the case of China, this is very obvious. So, with one little step, namely with this replica of the facade, we are positioning ourselves in three points: one is against originality and the typical world, one is a reply to the government, and one is a comment on the state of architecture today.

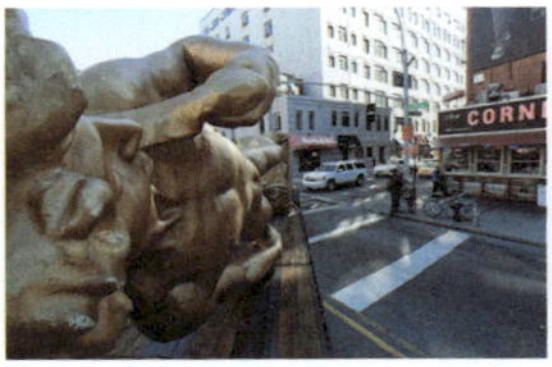

LET'S ASK THE QUESTION DIFFERENTLY: CAN YOU HAVE COPIES IF THERE'S NO ORIGINAL?

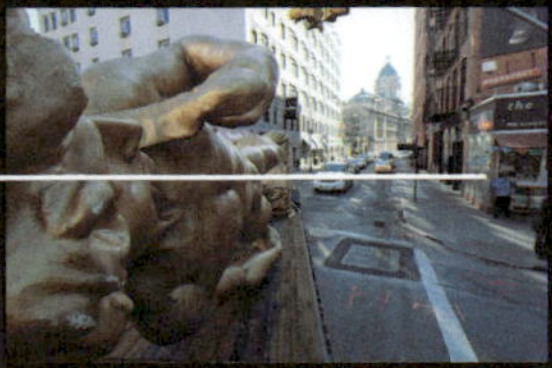

APRIL 06, 2012

Lynne Tillman, Cristina Goberna, Michael Morris, and Serkan Özkaya

- ● What interests us[1] the most is the replica and taking it in our own language, which is, the replica as a copy, but also the replica as an answer to the previous statement.
- ✦ Like "reply."
- ● Reply, exactly.
- ○ As in the Latin word?[2] That's interesting.
- ● Every time we build, we stylize, we rewrite. Basically this is what we do. We try to be systematic.
- ○ On Monday night, I was at the Guggenheim Museum for the Robert Rosenblum Memorial Lecture.[3] Rosenblum was an important art historian, and he died some years ago. The family and his widow started a lecture series. On Monday, Rosenblum's great friends Gilbert and George were guests that evening and were being interviewed by Michael Bracewell. And they all talked about "doubles." That was one of the most amazing performances I have ever seen; so elegant as well. Gilbert was born in the Italian countryside, and George is from the country in England. They met in 1967 in art school. What was amazing about their performance the other evening was that they had prepared the whole event as if they were one artist.

  Their writing is very much a part of their work. At certain moments, one of them (usually George), would say, "We have the text that we wrote in 1971 that we would like to read." So they interpreted these texts and

1 Fake Industries Architectural Agonism. See their manifesto, pp. 15–17.

2 Lat. replicare; to copy, replicate, reply.

3 Second Annual Robert Rosenblum Lecture: Gilbert & George in Conversation with Michael Bracewell. Guggenheim, New York (April 3, 2012). Since first meeting at Saint Martins School of Art, London, in 1967, Gilbert & George have created a body of art that describes with unwavering lucidity the experience of being alive in the modern urban world. Violent, confrontational, and intensely atmospheric, the art of Gilbert & George reveals what the artists have described as "the moral dimension" of their subject matter: the desires, fears, hopes, volatility, and frailty of the human condition in all its messy and democratic momentum. For nearly five decades, Gilbert & George have created their works entirely independent of any school or movement in modern and

read them out loud and at the end there it was – like a special tribute to Rosenblum. They did part of their singing sculpture. But it was just incredible. I had never taken them as seriously as I do now.

contemporary art, tirelessly refining the artistic capacities of what Robert Rosenblum so succinctly defined as "the singularity of their duality." They were joined in conversation by Michael Bracewell, a novelist and cultural historian and the author of several publications about their art.

⬢ I wanted to buy a ticket for that. It was difficult. We went to the exhibition they had at the Brooklyn Museum last year. It was really good. I didn't know they acted like that in talks.

⭘ They don't. It is what it is; they don't stop. This is their lives. I don't know how they are like in their home together, of course.

✦ The children don't want to see each other anymore.

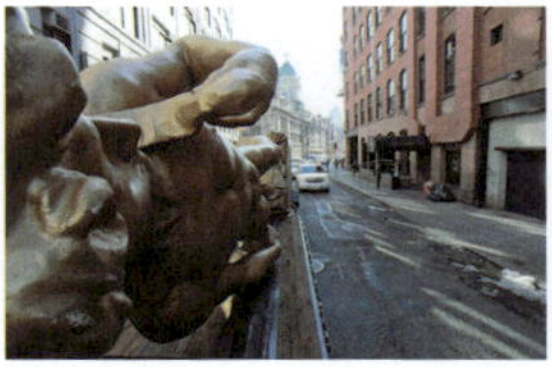

⬢ Is it because of what they do at the same time; that they act as one person?

⭘ That's right.

⬢ What we do as a couple, when we reflect on things – many times but not always – is that kind of arguing and disagreeing strongly on the state.

⭘ You are the opposite.

⬢ Yes.

✦ You talk? You fight?

⬢ Yes.

◈ Again there was a conversation with Gilbert and George, and I can see Gilbert rolling his eyes and getting annoyed when George is talking like, "I can't believe you said that." So they seem less than one.

⭘ I don't think Gilbert rolling his eyes is a

response of annoyance. Because he has this really odd gaze and he is aging quite differently from George. They are probably the same age, but they made this decision to be different from everybody. They wanted to bring emotion. They talk more about emotion, and I heard more from them about emotion than in previous years.

They talk about how they wanted to be feeling. They wanted to do everything opposite of what was going on in England at the time in late 60s. When they left St. Martins apparently they had a sculpture that everybody just hated.

✦ I was thinking about this other couple I had met in Amsterdam. No, I met them in Paris actually, but they are from Amsterdam, and I don't even know if they are still alive. They are two anorexic girls; super thin, almost dead, and they look like twins.[4]

Everybody called them twins. I'm not sure they were twins. I went to dinner with them and they would have shows. They are artists. The show would be a party like this, where they eat and drink, and then just leave everything, and make a video of this happening. They leave the video there with the leftovers to rot. That's the show.

They do a party in the gallery. They will videotape it, and leave everything as it is after the party and the video of the opening would be projected on the wall.

● It's interesting. It's about food.

4 L.A. Raeven. For the past ten years, twin sisters Lisbeth and Angelique Raeven have been working together as the Dutch artistic duo L.A. Raeven, creating controversial films and performances that challenge traditional representations of feminine beauty, anorexia and sibling relationships. Through a combination of home movies and excerpts from films and performances, the twins are observed in their daily lives, capturing the tensions, interactions and remarkable psychological connections between them.

✦ It's about food. And these people ate as if they were in a ritual. They ate at the same time, but how they ate was super slow. Once when we were having dinner together, one of them turned to me and said, "So are you working on a project?" I told her what I was working on. And then they kept eating and so forth. Ten minutes later she turned to me and asked, "Are you working on a project?" It was so scary. I think they had just lost it.

● So will you call this art? Or anti-relational?[5] Gilbert and George wanted to convince you that they are two people who are one artist. In other words, they are not saying they are one person, but they are the same artist.

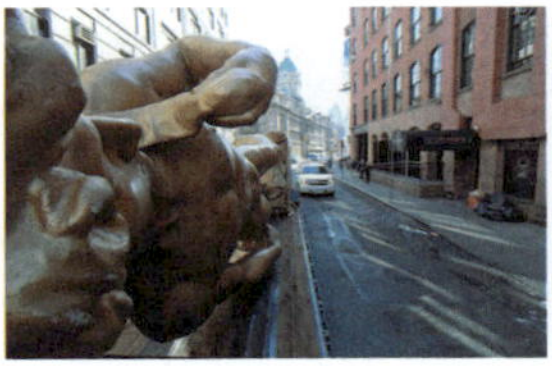

◈ What is interesting is that they can speak in a relationship because they are working as a collaborative. It's very hard for an artist to do that on his own. All that I am saying is that it is extremely difficult. Is it truly a solitary figure who talks about somebody else? At that point there is a dialogue, and then the emotions start to build up. I'm just saying it's very difficult. Do you ever enjoy people in art? I guess it depends on the collaboration. I think it's really important, and I think collaborative work is the only way to go.

● Is it the only way to work? Is it the only way to make art?

◈ Recently I worked on this sculpture – actually it's been about eight years – at the

5 Relational art or relational aesthetics is a mode in contemporary art, originally observed and highlighted by French art critic Nicolas Bourriaud. Bourriaud defined the approach simply as "a set of artistic practices which take as their theoretical and practical point of departure the whole of human relations and their social context, rather than an independent and private space."

Metropolitan. It was *Adam* by Tullio Lombardo [6] of the Lombardo family from Venice. It fell down and broke on the floor. It was then stored there for two years. They didn't know how to go about this project. So after two years, they asked me if I could have a look and talk with them about it. And then they hired me on contract.

- ⭘ You were a sculptor?
- ◈ In the past I worked in this way.
- ⭘ For what did they hire you?
- ⬢ When did they hire you? Did they restore it before?
- ◈ Well they'd done a little bit. They hired me based on what I'd done in restoration and conservation. It didn't have anything to do with what I've done previously as an artist, which I stopped doing around 1995.
- ⬢ Another question. Is there any precedence for this work? Or is it something completely new? What I remember reading is that the sculpture broke down in many pieces. Is there a regular procedure that museums follow in these cases? Or is this something new? Are they trying to restore it perfectly?
- ◈ Yes. Perfectly.
- ⬢ From the pieces? From the original pieces?
- ✦ What ignorance! That you think you can take it back. Something is broken.
- ⭘ The Japanese and Chinese model...
- ◈ In the end, it will be all put together and it won't be obvious that it had been broken.
- ⬢ You won't see it?

6
Tullio Lombardo (1460–1532) was an Italian Renaissance sculptor. In 2002, Lombardo's marble statue *Adam*, part of the Metropolitan Museum of Art's collection, fell to the floor and shattered. The sculpture, created in the Renaissance, was the first recorded free-standing male nude since antiquity.

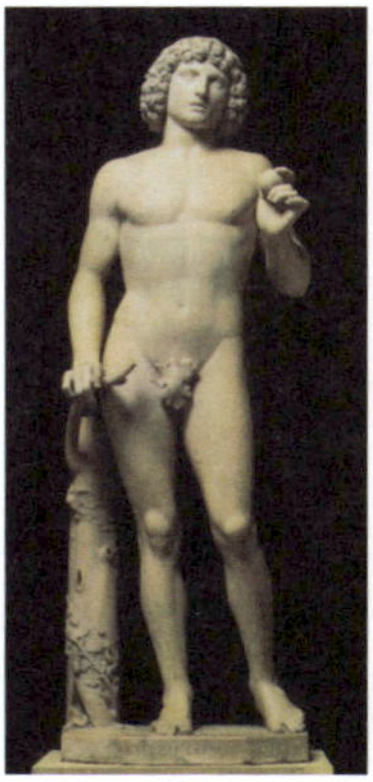

◈ I don't think you'll see it.

● What is the value of restoring it exactly with the same pieces?

◈ What options do you have? You have all these broken pieces there on the ground, you pick them up, and transfer them to a place to put them back together. You spend a lot of time doing it. Basically there are some areas, mostly in the arms, where the fracture is so catastrophic that there is no way, no possible way, to put these pieces back together. But out of five tables of broken fragments, we are down to half a table.

○ My God. How many people do you work with? That's what I wonder.

◈ There are really just two of us, but there is a collaborative team of seven or eight people supporting the work. And the museum has really done a great job. They have all the pieces scanned into CAD, then carved with a CNC machine. I have a 1:1 model that I can work with. I don't have to work with the original. If I have any issues of how to work the pieces, I use this copy. I never have to touch the original. In many ways if you keep handling these pieces they just get worse and worse, so you try to handle the original as little as possible.

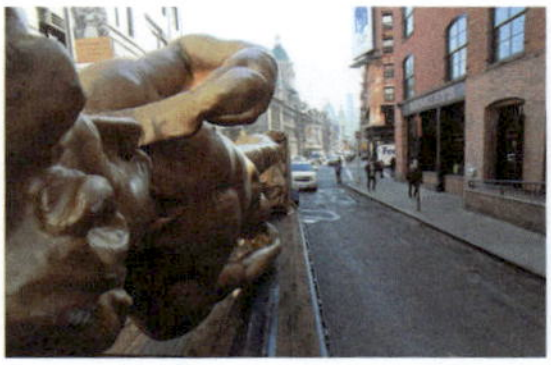

✦ So you fix the copy first?

◈ I fix the copy first.

✦ How did it go? Is it better than the original?

◈ No, actually it's worse, because…

✦ Who gets to keep the copy? At the end?

◈ The copy will never go together. The copy is only used as a tool that helps to put the original back together.

✦ That's what you think. Maybe it's more valuable than the original.

◈ No, trust me; it's not. I can't imagine it will be. It might be of some benefit. But I mean it is a piece of plastic.

- How do you feel about doing this work?
- Mixed. There was an opportunity to buy a figure that looked like *David*, but in a very odd way. It was about six feet tall and made out of marble. Chinese marble. It was on a website called: *www.wishihadthat.com.* And we had this meeting with about six to seven people. The curators asked me if the figure would be beneficial as a model to work with. We already had a 1:1 and that seemed good enough to me, but here was this marble. I did't think we needed it and it was a $7,000 expense. Nonetheless, I was outvoted. Everybody wanted it, so the marble was purchased. The *David* came to the museum, and we proceeded to break it apart exactly along the lines *Adam* was broken.
- How can you do that?
- There is a way of splitting the stone – the ancient method with bone, feather, and wedges. You drill a hole in, you put a wedge in with a feather, and you tap it. If you tap it in two or three holes, you propagate the crack and you can break it very precisely along these lines. So, we broke the sculpture apart and we used it to establish the methodology for putting the original guy together. We worked from two different copies.
- From marble as well?
- In the fact it was marble too.
- And the color at the end? They were two different materials?
- We are using the marble, Chinese marble, Sichuan marble, hideous marble, and then the plastic model. Then when we get to the original, we are putting that back together.
- It was only to work for the copy? You broke the sculpture only to work with the copy?
- Only to work with the copy.
- You were totally right.
- So you broke it and glued it back together.
- Was it old?

◈ No, it was a contemporary piece. We devised the armatures to put the original back together without any pins or adhesive. We put together the whole sculpture – the original – without pins, so that it was standing.

● That's extraordinary.

◈ In the end we will use adhesive and pins.

✦ The whole idea is to have adhesive that is reversible, so that if in the future they actually come up with a better way to put it back together, they are going dissemble it and reassemble it.

● Is this a fifteenth century or a sixteenth century piece?

◈ Yeah, it was done prior to *David*, most likely Michelangelo saw it prior doing his *David*.

✦ Same posture, same thing.

◈ 1495 or 1496.

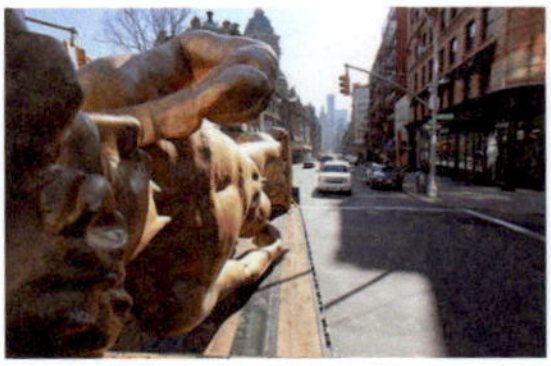

✦ Oh, that close to *David*?

◈ I think *David* is thirty years later.

✦ No, it's from the beginning of the sixteenth century.

● How long have you been working on restoring this statue?

◈ Maybe six or seven years.

● Do you go to the Met every day?

◈ Not every day. When I'm there, I spend about three or four hours a day on the work. Then, I'm diverted off to other projects – like this Michelangelo piece.

● Where is this?

◈ At the Met. Twenty years ago, Kathleen Brandt, an art historian at NYU, noticed this figure. It was on Fifth Avenue and 79th Street in the French Consulate building, or something like that, right on the corner there. Stanford White had used the figure in a fountain. He had purchased it in London around 1902 and put it on an odd pedestal, one which was actually designed for a candelabra. The entire piece became a real pastiche of things. White had the Piccirilli Brothers in Brooklyn[7] carve turtles into the work and make this huge fountain as part of the entrance. Anyhow, Kathleen walked by this figure they were filming and she said, "This is something done by someone that is very, very familiar to me." And she thought it was by Michelangelo. For the last twenty years, the Met has been trying to figure if her assessment is correct. They have put together a record from Michelangelo's students and found a drawing of the figure rendered in his hand. They found another drawing from maybe fifty years later.

✦ Was the drawing full? So you know what the legs looked like?

◈ Yes. And so, yes, it's broken. I think people accept it as a Michelangelo; done when he was sixteen. It was most likely his first piece.

● So, Stanford White broke the legs so he could...

◈ No, but the part of the story that I like is that a sixteen-year-old is not going to be able to get a piece of marble to carve. He is too young. They gave him a piece of stone – an ancient stone, a Roman column. It was an impossible piece to carve. They gave it to him and said: "Try your hand at this." And he proceeded to carve this *Cupid* or *Young Archer*. Now we've removed it from that pedestal and put it on another pedestal, which is sort of anonymous and circular like a column.

● Like a plinth?

◈ Like a plinth, but…

7 The Piccirilli Brothers were a family of renowned marble carvers and sculptors who carved a large number of the most significant marble sculptures in the United States, including Daniel Chester French's colossal *Abraham Lincoln* (1920) in the *Lincoln Memorial*, Washington, D.C.

✦ Very modern.

◈ Yes, very modern. And then we changed the moment of the sculpture. Because as it was sitting, it was sort of leaning forward and off to the side. Everyone wanted to bring it back. There's some reference to Michelangelo's teacher, Bertoldo,[8] I think, but this I don't understand. When you look at the figure there's a twist, but the chest is facing forward. Apparently, the pose is a reference to Michelangelo's teacher. We twisted it a bit. Actually it came back; it went to the side a bit, the proper left. And then the position of it in the gallery is sort of direct. You see the twist in the body, but the chest is facing forward. That piece went up yesterday. It's in the Renaissance gallery.

● So your work was?

◈ To change the position of the piece and mount it on a new pedestal.

● And it didn't have legs?

◈ And it did not have legs.

✦ You should put a pair of legs on.

◈ We added these supports that you don't like.

✦ Because they put them low, as if that's the real height of the statue.

◈ Now see, that's a good point. You know, once you elevate it then there are all sorts of issues. How do you complete the bottom of that? But the idea was to elevate the figure so that one would get the sense of how it would stand.

8 Bertoldo di Giovanni (1435/40–1491). Born in Florence, he was a pupil of Donatello, and for a long time worked in his master's workshop. He later became head and teacher of the school for painters and in particular for sculptors—a school which Lorenzo de' Medici had founded in his garden. Some of the most significant sculptors of their time attended this school.

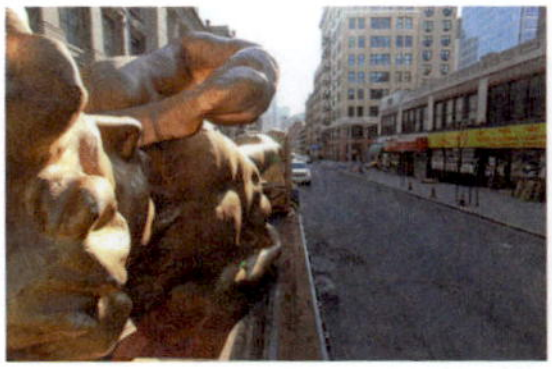

- ⭘ Do you think it should just be floating with wires?
- ✦ Not with wires, but maybe with some…
- ⭘ No, I mean if you hung the statue from the ceiling?
- ✦ Yeah, you could… Or maybe you could use poles.
- ◈ Now it just comes down to the knees, and of course, they are oddly broken. Since it had to sit on the pedestal, we made these small epoxy inserts that were cast into the bottom of the fractured surface so they click into the bottom of the break and sit flat on the pedestal. It was a bit of a trick to do that, because it's basically standing on these two little points. This thing is heavy.
- ⬢ I guess the other way would have been to fix it to the wall or something like that.
- ⭘ But then you couldn't walk around it.
- ◈ Everything that we do can be taken apart. If they don't like it they can do it again. And they will; sometime in the future, somebody will. In the spirit of copies, the museum is making copies of this with the pedestal to sell in the gift shop.
- ✦ Small copies?
- ◈ I don't know how big.
- ⬢ Regarding the other sculpture – the one that is taking so long – do you think that a big museum such as the Met is putting in such a huge effort because we are here in New York or in America in general? I can't imagine that happening, for example, in Europe. I come from the south of Europe and I'm an architect, and any time you are trying to make a building there, you find things like this fairly regularly where you don't know the author.
- ◈ Right. But do you try to preserve them?
- ⬢ It's a big problem because when you make a building, you can't really touch what you find. Normally, you would find columns or pieces of streets. You have to leave them in place. You can't really touch them. Sometimes people will hide them.

People might want to show them in the ground floor, and you have to change the design.

◈ We basically just reversed everything out. There was a brass pin inside the figure. We just took everything out.

● Is that what Stanford White did? Put in a brass pin?

◈ Probably. He put a piece of copper plumbing pipe in one side and the pin in the other.

● Do you think he had any idea what he was working with?

◈ I don't think he did it. Someone else did.

● He didn't know that it was a Michelangelo or that it might be?

◈ It was sold as a Michelangelo, but nobody believed it. That's what I learned today. The Met doesn't own this piece. The French government owns it. It will probably go back to the Louvre after eight years or something like that.

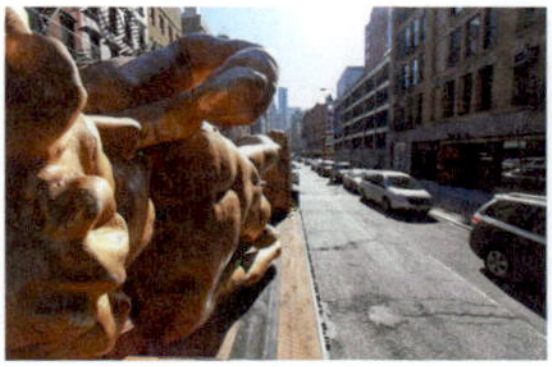

● So who paid for the fixing of it?

◈ The Met.

● And yet the Louvre owns it.

◈ Yes, I think that's it. But in terms of copies, they crank these things out. It's this huge moneymaker. I go into the museum bookstore now and it's just a disaster. They have note cards filling up the shelves where books were.

⬢ This is happening everywhere.

◈ It drives me mad.

⬢ You go to the Strand bookstore, or you go to Apparel in Columbia University, and you can buy soap.

◈ Exactly.

✦ But you need those things. Soap. It's essential. How would you bathe without soap?

● What are you going to do to advertise your *David*?

✦ They commissioned a necklace with charms. Two little charms. One is *David*'s penis and one is the fig leaf. (Img. 1)

- I thought you were going to say *David*'s penis and then I thought, "No Lynne, you're just being ridiculous."
- No, and I really like the owner of the museum[9]. I think he is a visionary. He just mentioned it in *The New York Times* article and now it's all over the place. Everybody is asking about that necklace. Everybody! These serious journalists from Argentina who are interviewing me about my work ask as the last question: "So what about the necklace?"

9 Steve Wilson, founder of 21c Museum Hotels.

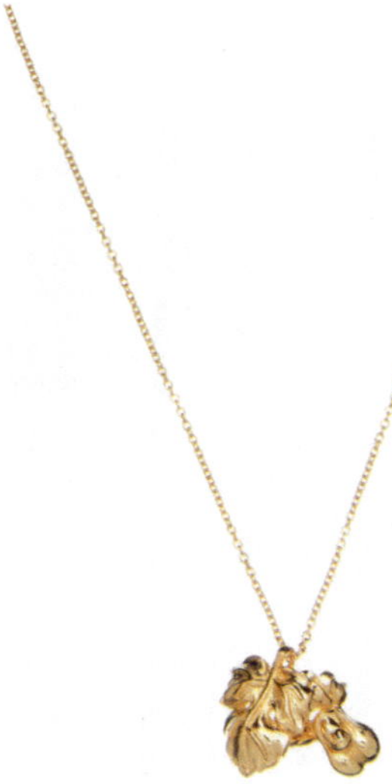

Img. 1

- Really? This is so interesting. Do you remember from the conversation at Storefront we were talking about how the fake is an

original too?

◈ I remember seeing one of Andy Warhol's doubles. Nobody knew it was him. We were in Oregon. We didn't know who Andy Warhol was as a pop artist and exactly what he looked like.

⬢ His doubles didn't look like him at all. You do a copy of something, and the copy of the copy. Maybe in the end it is not a copy any longer.

✦ To start with, copying is the actual act that entitles something as the original.

⬢ I'm not so sure. You have originals without copies; like you, for example. Do you have a copy? Maybe you have it and we don't know.

✦ What makes you think I'm so original?

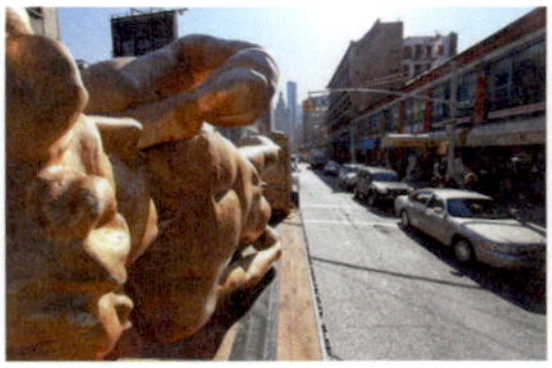

◈ I think I told you. When I went to the Parthenon we had Italian mold makers to cast and make molds of the reliefs on the west facade. We did the *Orion* in the west facade; the rest was taken down for some reason. The walls were made first and then the molds were shipped out to London. When the facade came down, a lot of it was broken up. So the plaster was in many ways more profound than the original in terms of aesthetics, because a lot of stuff has been broken up in the process of moving it.

✦ Do you think it's more valuable than the real piece? Because it's all about the value.

⦿ So you are saying that there is no such thing as an original? Linguistically, that's true. There is no such thing as "less valuable." But when you say something is original because there's a copy thereof…

✦ We call a thing original when there are copies of it. That's what I think. Your Louis Vuitton bag is an original, because there is a bootleg copy of it. It will not be original by itself. Michelangelo's *David* is the original. Linguistically, a copy needs an original, and a copy cannot stand by itself, nor can the original stand without a copy – it's a symbiotic relationship.

⭘ No, no, no. Linguistically, if you say something is original, you mean it is the first. You don't necessarily…

⬢ The only one also…

⭘ This is the original manuscript.

✦ And then this is the printed copy that I bought today.

⭘ It has nothing to do with the meaning. It is a book; it has nothing to do with a copy of the manuscript.

✦ I know. Is this the thing that you have written now? Take a music piece. Where is the original? Is it the music in the air? Is it the version that the composer conducts? Is it the manuscript or the printed version of it? Is it the idea of music he had in his head? One could keep going back forever.

⭘ I think you are mixing the numbers. I understand why you want to do that, but I still think there is a logical problem. I'm just not sure how to get to it. What about origin? You are saying that "original" and "origin" are the same thing. Does anything have an origin? Can a thing have an origin?

✦ In order to have an origin, you need to have the thing that is a derivation of the original. Can you give me an example of origin?

⭘ The clay; the origin of the clay was Ethiopia. This is Ethiopian clay.

◈ I don't understand. Sorry.

⭘ This clay comes from Ethiopian soil. That's the origin of this material. That's an origin.

✦ And your book comes from?

⭘ What I am saying is that the particular substance is original.

◈ I think what you are saying is that everything is original, even the copy. So, every piece of clay you grab from that ground in Ethiopia is a different piece of clay.

⭘ It is still from the same area.

◈ It is from the same area, but every piece is different.
● That doesn't mean it is original.
⬢ I think some parts make sense. The origin, physically speaking, is not an original if you don't have a place; if you don't move things to other places.
● It doesn't really matter where it is from if you never moved it.
⬢ Exactly.
◈ You make this "origin" sound as if it's static – like it is a place.
● Let's say the origin of my family – as far as I know – is such and such a place…
✦ A family is a tricky example. You have to go back and back and back and back.
● In our case, we don't know. After a few couple of generations – mud, mud! We came from mud.
✦ Ethiopia?
● Exactly. In fact, that's what Adam is: man of clay. How did we get there? Now, talk about origin stories!

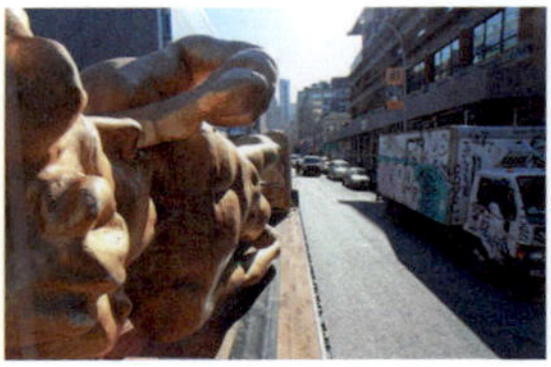

✦ In fact, what you were saying is that every single thing is original and that is valid. Of course, this is original, that is original, this is original, that is original. Although you have another…
● But if you say everything is original, then original has no meaning.
⬢ I don't think that everything is original.
✦ But everything is here and now.
● I don't think I was saying that… It's a very interesting point.
◈ I was just trying to take it…
● … to a different place.
◈ I don't like to see origin as a static thing.
● Why not?
◈ Because, I guess deep down inside, I really feel like everything is original. I do. If you consider these two glasses together, they might be of the same material, but they might also be off ever

so slightly in terms of their respective material compositions.

⭘ So what's at stake? Let's make a list…

◈ I'm just saying that it is different. We may not perceive it.

⬢ It depends on the grounds that we are talking about. Are we talking about the form? Are we talking about the specific glass? This glass is different than the other one.

✦ And they also occupy different spaces in the universe.

⭘ Let's ask the question differently: Can you have copies if there's no original?

✦ I would like to think so. There's me, and you, and you…

⬢ We would like to think so. Can we make copies without an original? Yeah…

⭘ What is a copy then? Everything is original when there are no copies.

✦ The original is a non-existent thing. The original is god and we are examples of god.

⭘ There is no god.

✦ Yeah, exactly. That is what I'm saying.

◈ I would disagree with you. I think that there are originals. And I think there are many copies. I would go that far. It flips for me.

⭘ Because, you're saying, in the making of anything it's the first time it is being made.

◈ It's a fingerprint of that thing that's made.

✦ So what do you think about god though?

◈ Oh, I don't have a thought about that.

✦ Do you look at it as original?

◈ As the copy of an original?

✦ The absence or the presence of god? Is that an original or a copy? You were saying everything is original… Everything? Yes. We understand that this is original; that that is original. What about an abstract concept?

⬢ I guess it depends on the religion that we are talking about.

⭘ So if you believe everything is reincarnated…

✦ Reincarnation? Now we went in a different direction.

○ Do you think about god?

✦ In the Old Testament, Moses is the only one who can hear god, and he tells his brother. And the brother becomes his mouth, and he speaks to the people. In that respect, I was trying to trace it back to god.

Also, we are examples… We are made after god, but we are imperfect, mortal, and so on. We are real and god is unreal, yet we are imperfect and he is perfect. QED! There is no god. No one has seen it. No one has heard it – except for Moses.

○ What about Christ?

✦ Yeah.

○ That's an answer, "Yeah."

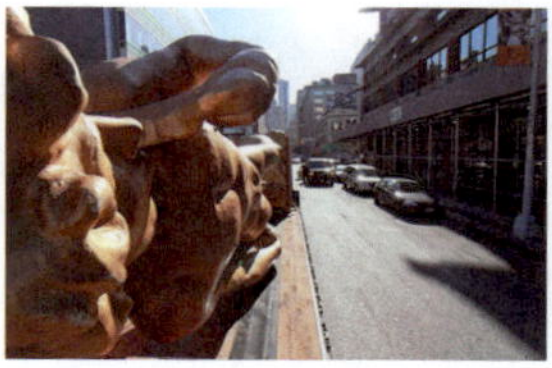

⬢ Some people apparently saw him and he is the son of god.

✦ Christ? Well, son of god doesn't mean… You know the father doesn't really count…

○ In this novel I'm working on now my protagonist is a cultural anthropologist who's involved in photography. He's studying family photographs and also studying his male peer group.

✦ Zeke?

○ Yes, Zeke. He's about thirty-seven and he is interested in how his age group thinks about women – and their attitudes towards women, post-women's movement and so on. It's given me a chance to work through some stuff. But, in one of his many thoughts, he is thinking about why the idea of monotheism was a sign of a higher civilization. And this is something that, in terms of the copy and the original…

✦ That's nonsense. That's the decline.

○ Well, it's interesting. That's what I think, but why get rid of all these great mythic figures and these different deities and leave just the one? But see, you couldn't have Western Art without that, could you?

✦ It's true.

⬢ But also, consider what happened with the Holy Roman Empire and their treatment of Christians at first and then later on of peoples throughout Europe. They were barbarians.

⭘ But in terms of the emphasis on the original, if you get rid of multiple gods…

⬢ But that's multiplicity. It's different. All the gods from Greek culture, they were all originals in a way. There was Zeus of course, but all of them were similar gods. Not like multiple theoi, multiple originals.

⭘ But, okay.

◈ They could take different shapes, too.

⬢ Yes, they could modify their shape and become human.

⭘ But there is some way in which the emphasis on the one as being… Singularity becomes a huge issue in Western Renaissance.

✦ Monotheism is in our lives. And it's very totalitarian. It divides everything into two: believers and non-believers, for example.

⬢ Yes, but at the same time, only in the basic principle it is only one. If you go a little bit further from that, it's really three. It's the holy spirit, the son, and the god. And then there's a huge multiplicity stemming from that – every neighborhood has a saint.

✦ But god is one, and always male.

⬢ In Greek culture it was Theos – it was male. It was one and then there was the multiplicity of the saints.

⭘ I was looking for female myths…

⬢ Christian?

⭘ No, not Christian. Pre-Christian. There were a lot of sons and fathers and I was looking – this was years ago, while working on another book – for mothers and daughters, and all I could come up with was Demeter and Persephone. You've got Medea who kills her children, but it's not really a mother-daughter story. I was trying to see where the parallels were. But this is going far afield of…

✦ No.

⬢ No, I mean in Greek culture there are many female figures.

⭘ Yes, there are female figures, but there are not myths about mothers and daughters. I was interested in the struggle between mothers and daughters. You have, of course, the tragedies like Medea or the daughter Antigone who wants to bury her father. But there's really very little about mothers and daughters.

✦ Well the rumor has it that…

⭘ Rumor has it?

✦ … that in the very old days when people had no idea of the connection between men and childbirth, it was a maternal society and women were superior.

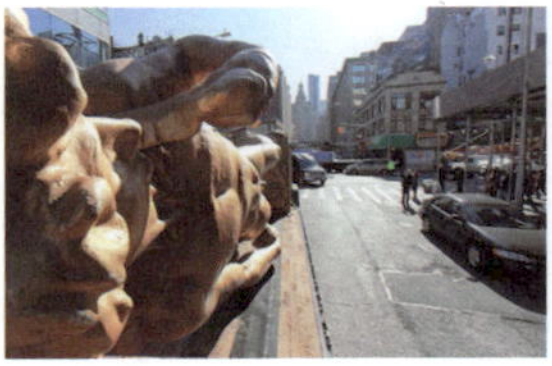

⭘ Yes, the matriarchal.

✦ Matriarchal. And the men were slaves or workers, and if a man was ambitious or greedy enough to elevate himself to the level of women, he would start carving stuff like making sculptures or being an artist. The artist would work to create something which is not a human being, but rather something that resembles it.

⬢ Copies!

✦ Copies. And in making these copies, man would go just one level up. And that's the birth of art – at least according to one theory. And once they discovered the correlation between men and having babies…

⬢ … that was it.

✦ Yeah, everything went downhill from then on.

⭘ Years ago, around 1974, I had a conversation with George

Maciunas,[10] one of the founders of Fluxus. He was crazy; sort of a mad man.

- ⬢ All the time? Or only when he was performing?
- **o** No, he was just… He was out there. But he was also very sharp and clever. For him the question of who did something first was very important,. And I remember saying to him, "What difference does it make?" And he said, "It makes a difference who did it first." So even though he was the progenitor of Fluxus, which was a correspondence school of Art, he still valued the idea of who did it first. That's basically what you want to knock down. It's not so much origins as the importance of the origin. It's different, isn't it?
- ✦ Forget about the hierarchy between the original and everything else, or who came first and who came second. Then, you also imply that who came first didn't actually matter.
- **o** Not mattering is different from whether or not it is happening.
- ✦ Do you think?
- **o** Yes, I think to say that it really doesn't matter who… Okay, you could say that a number of different people could be credited with developing photography. I took a course in the history of photography with a great photographic historian. Why am I forgetting his name? Anyway, he talked about the origins of photography as being multiple. There were many different people working on it from the late eighteenth century on. And this was a development for positivism in terms of getting

10
George Maciunas (1931 – 1978): Founding member and central coordinator of Fluxus. He is most famous for organizing and performing early happenings and for assembling a series of highly influential artists' multiples.

an imprint of an image. We use the date 1839. (Img. 2) In fact there were three main figures in 1839 that were all doing the same thing. It doesn't really matter who invented photography.

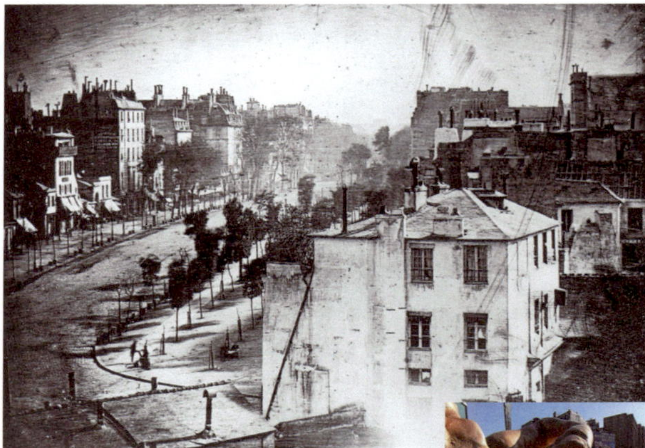

Img. 2

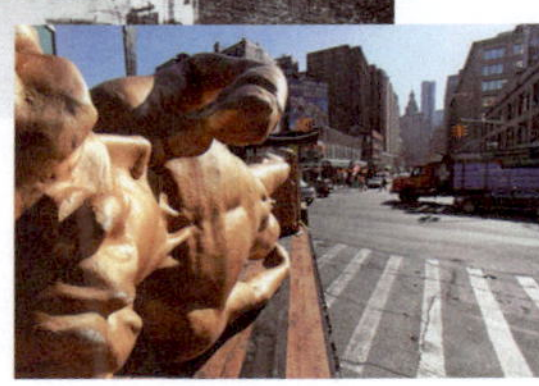

✦ It does or it doesn't?

● It doesn't, but there were origins that took it forward.

✦ Even according to my experience, I know that in the arts different people come up with similar ideas at the same time. At the end of the day, the more obsessive one or the more famous one or more acknowledged one always…

● … gets the credit!

✦ Yes, he always gets the credit. People forget about the other ones.

● The other one has more infrastructure.

✦ Sometimes.

● More medium in something; not necessarily the original idea.

✦ Yes, but sometimes people come up with great ideas, but they don't cling to it. They don't really claim ownership. They are not greedy. There are other people who are obsessed with that thing, and they feel they are the creator of that, and they get the credit, which is, I think, fair enough.

● The reputation of the copy is something we think about a lot. In architecture, the replica is terrible.

◈ In terms of? Terrible because?

● I guess it's because architects are absolutely obsessed with originality.

✦ So are the artists.

● Yes, but in art there have been many experiments and many people working with copies; with many different types of copies. Now, we are trying to do the same in architecture. So, in art, irrespective of the reputation of copies or replicas, as a practice it's something that is well known. In architecture it's not at all like that. So, even though we both copy consciously – although I understand the difference between copying and simply having references – we cannot come out and say what we're doing. You *can*, for example, speak about the *Barcelona Pavilion* by Mies van der Rohe…[11] (Img. 3)

○ And that was where?

● In Barcelona, and it was built for the universal exhibition. It was demolished and then copies were made at two separate times. But if you look deeply enough, you see that the copies were made with certain technological advances. The copies were not built in the same way. They were never able to build the same.

11 Barcelona Pavilion or German Pavilion for the 1929 International Exposition in Barcelona, known for its simple form and its spectacular use of extravagant materials. Iconic furniture was designed specifically for the building, e.g. "Barcelona Chair."

12 Refers to Roland Barthes's essay, *The Death of the Author* (1968), and Michel Foucault's, *What is an Author?* (1969).

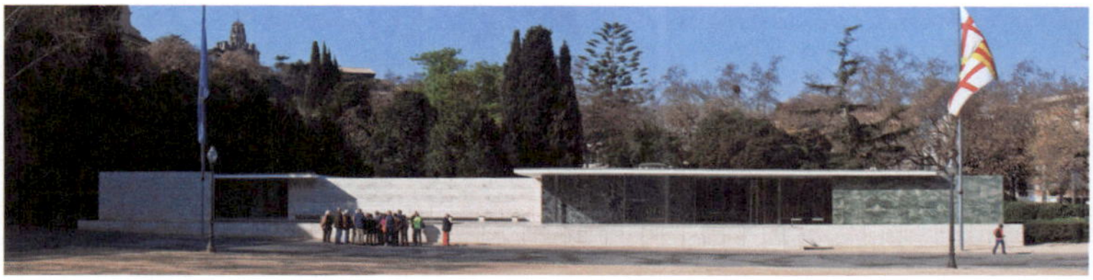

Img. 3

○ Could you say that the design was the same?

⬢ If you see the plans and you don't know or look very deeply, then yes, they look the same. But then again, if you see a structural method, for example, it doesn't follow that the implementation will be uniform across all the resulting buildings.

✦ Maybe you are looking too close?

⬢ Maybe.

◈ You can't look close enough.

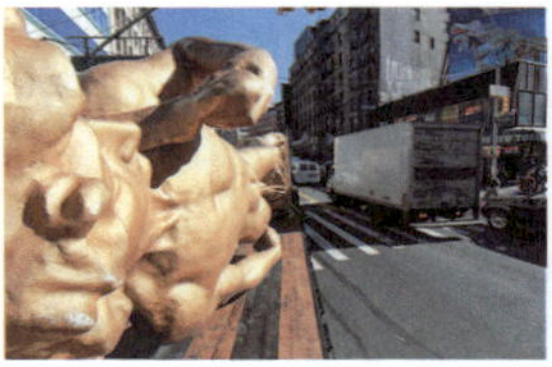

⬢ We see this done all the time. And you say that in art the reputation is terrible?

○ What reputation?

✦ There is this obsession, and people are so selfish and narcissistic, per se, that they wouldn't actually accept another person's idea. I remember when we were trying to have this MFA show at Bard College. We were students, not mature artists, and we were trying to find our voices. We were trying to graduate and wanted to design an announcement for the opening, and every time somebody gave an idea, "Let's do this, do that…" etc. somebody else would say, "Oh, but that would be your idea!"

Then we ended up having this blank page with just the names on it. People are so obsessed with originality. They naively believe that originality comes from within yourself – that meaning comes from your body and mind.

○ They haven't read Barthes or Foucault.[12] That's their problem.

◈ What? I'm sorry?
○ Barthes or Foucault.
● What about copies in literature though?
○ But you see, Virginia Woolf said a great thing. She said that books continue each other. But in art, it's true also. You wouldn't have Jeff Koons without Andy Warhol. Warhol wouldn't exist without Duchamp.
● These are references. You'll always have a million references – all the books that you've read, all the architecture that you've seen, and have studied. And then you produce something. But it is another thing if it's a replica or a copy.
○ Do you mean plagiarize?
✦ Yes, like Lautréamont?
○ Do you want to explain?
✦ He composed texts from other texts,[13] but that was almost 150 years ago, so it was a kind of blasphemy. I really haven't seen such a thing today – practiced so naively, without any other layer – except for the *David* statue, which is twice as life.
◈ You're talking about plagiarizing in art?
✦ Yes, well…
○ No. Plagiarizing would mean that you had taken something.
◈ In the 1980s, there was this guy, and I can't remember his name, who was making these Duchamps.[14]
✦ Yes, but in his own way. The drawings, I know…
◈ No, no, no. Not in his own way.
○ You don't mean Robert Gober?

13 "In spite of his obvious application of this method [plagiarism] to theoretical language in *Poésies*—where Lautréamont (drawing particularly on the maxims of Pascal and Vauvenargues) strives to reduce the argument, through successive concentrations, to maxims alone—a certain Viroux caused considerable astonishment three or four years ago by conclusively demonstrating that Maldoror is one vast détournement of Buffon and other works of natural history, among other things. … A slogan like 'Plagiarism is necessary, progress implies it' is still as poorly understood, and for the same reasons, as the famous phrase about the poetry that 'must be made by all'." Guy Debord and Gil J. Wolman, "A User's Guide to Détournement," *Les Lèvres Nues* (1956).

14 Refers to Mike Bidlo, who produced accurate replication of the works of significant modernist artists, including Picasso, Matisse, Duchamp, Warhol,

◈ No, I can't remember his name. He did this stool with a bicycle wheel…

✦ Yes, but he went through a Duchamp era; he went through a Picasso era. But he also…

◈ It's a kind of plagiarism.

✦ He played with that idea.

⭘ But that was a part of appropriation.

✦ Appropriation, yes. Plagiarism doesn't really exist as a form of art. You're right there.

⭘ Except for the guy who did Vermeer.[15] He said that his paintings were Vermeers. Remember? There was a great scandal. There were paintings that were so great that they were accepted as Vermeers. They were copies.

✦ Upwards of twenty percent of the things in museums are considered to be fake.

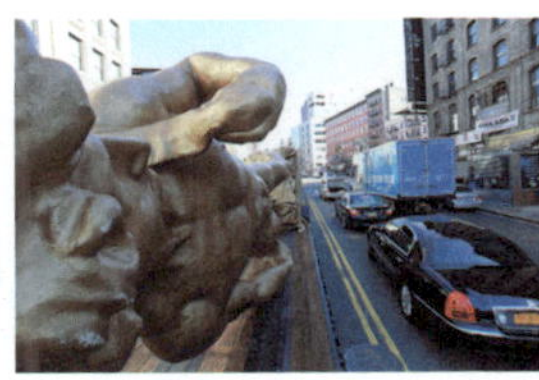

⭘ What we're talking about, though, is copying and not plagiarizing in the sense that they try to pass off the work under their own name – it's making something under somebody else's name, because it's valuable.

⬢ That's another issue.

✦ That's another intention, though. It's trying to make money by selling it.

⭘ Yes, but nonetheless you could say that this is a great painter. He could make something look like… He didn't have the conception of it. He was following. So that's a different thing. What about the conception? What about the conceiving of something? In writing, several years ago, I came up with the

and Brâncuşi. In 1984, he recreated Warhol's *Factory*, and in 1988, he exhibited 80 copies of Picasso's paintings of women. Bidlo's approach has caused outrage amongst some critics and has gained him some notoriety.

15 Refers to Han van Meegeren (1889–1947), a Dutch painter and portraitist considered to be one of the most ingenious art forgers of the 20th century.

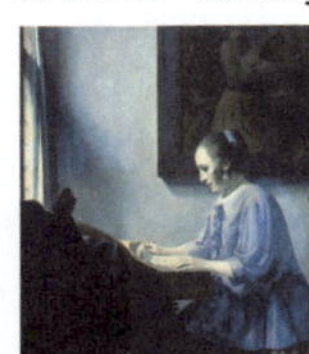

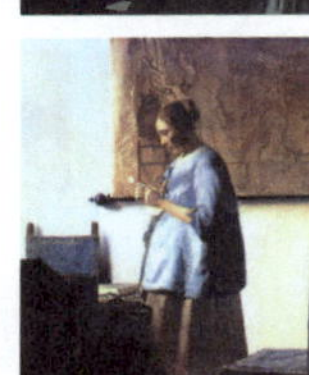

character *Madam Realism*. So what should I feel (it hasn't happened… well actually it did happen) if somebody writes something using my character, *Madam Realism*?

● The name or the character?

○ Well, the name. But the name and the character are…

● Wow, incredible.

○ Yes, it did happen once. But it's no big deal because I'm no big deal, so it doesn't really matter. It's a very strange… It's like having your checks stolen.

● And did the character behave as you would have had her?

✦ Even better.

○ Yes, it was much better than anything I could have written.

◈ That's like with Borges.

✦ When they asked Eco how he felt about the English translation? He said, "Oh it's much better than the original Italian!"[16]

○ In one of my early works, a German translator said to a friend of his (who told me), "This is okay, but I could make it really good in German." But he didn't. He didn't translate it, unfortunately. Well, that was a tradition in Western literature.

When you're – and this is just speculation – breaking away from the church in those early moments, you still think there is something coming from something else.

✦ Borges is a great example of collaboration. He wrote with his collaborator[17] – I don't

16 On William Weaver's translation of one of his books.

17 Refers to Adolfo Bioy Casares (1914–1999). Their collaborative works include *Six Problems for Don Isidro Parodi, Two Memorable Fantasies, A Model for Death, Chronicles of Bustos Domecq, and New Stories by Bustos Domecq*.

18 Dr. Marc Levoy.

19 *Digital Michelangelo Project*. An outline and a smaller version of the model can be found on the website: http://graphics.stanford.edu/projects/mich/

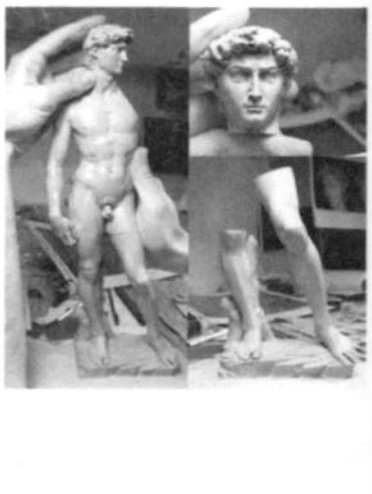

remember his name, so maybe that is a similar example of whoever is more famous gets all the credit – but then they used to write all these books together as one author. And then he used to add layers and layers and layers of phrases such as: "I took this from this encyclopedia," or "I wrote it with someone else here," etc. ...

○ Let me ask you a direct question: Why did you do the *David*?

✦ Why? There are a couple of reasons. First of all...

◈ First one filled in, right?

✦ No. Why the first one?

○ Why did you have the idea to copy the *David*?

✦ Precisely because it is the symbol of Western art, or Western sculpture, but also that it was available as a computer file, in the format of a three-dimensional file. It had been scanned by a Stanford professor[18] years ago and I bumped into it on the internet.[19]

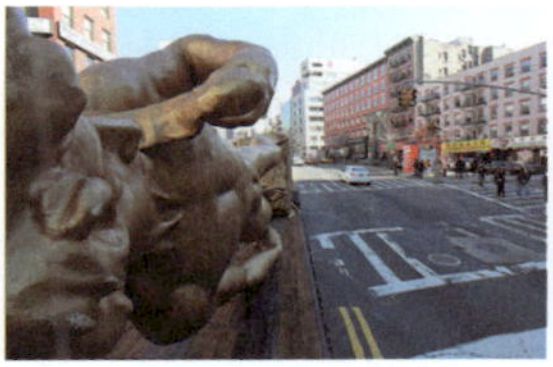

● Really? You did it from an internet file?

✦ No, I got it from him, but in order to get it I had to get permission from the Italian Cultural Ministry and Stanford University and so on. My reasons have nothing to do with *David*, nothing to do with Michelangelo, nothing to do with the form. If we had accepted Donald Judd's boxes as a symbol of our civilization's sculp-tural symbol, then I would...

◈ ... and you had the computer file...

✦ That wouldn't be too hard... So, I think that's probably why...

○ Do you think you are basking in his aura?

✦ Aura?

○ Works of art have a kind of aura.

◈ But, do you see it as *David*?

○ I haven't seen it.

◈ I have seen it in New Jersey and I don't see it as *David*.

✦ You haven't seen it in New York… sideways… and…

◈ … and that made a difference?

✦ It made a huge difference. I am not kidding you! It had a huge presence.

◈ It might have a huge presence, but the question is: Do you see it as *David*?

⬣ There is a difference, of course. It's golden and so on. But it doesn't matter to me. It is so impressive.

◈ I am not saying it isn't impressive or less impressive. I give sermon that it is impressive! It's just not *David*. I don't think it is even a copy. That is why I'm going back to my original premise; this is an original.

✦ To be honest with you, I haven't seen Michelangelo's.

⭘ That's because you are perverse. I mean that is a really perverse thing to do. I am going to go tell the press that.

⬣ What are you talking about?

⭘ Everybody does this, and in writing also, when they say, "your book is so much like blah blah's." I respond, "I have never read that book!" You've seen pictures of the *David*?

✦ Oh yeah! I know what it is.

◈ The double part makes sense because it's twice the size. I can understand that—in that way—using double. But in a scale there's everything, and for me, twice life -size just doesn't fit in that package. Do I recognize it as *David*? Of course I recognize it as *David*. Does it seem that it is a copy of him? No. For me it is not a copy of *David*. That's okay; I have always thought of it as your work. I never thought it was Michelangelo's work.

⬣ I think it depends on the boundaries with which we define copying.

✦ People also steal. With today's artists, people see copying as a way of making money and they steal.

⭘ But that's been going on forever.

✦ Do you think?

● Of course.

✦ I don't think so.

● Isn't that the idea with Picasso, who said, "Great artists…

✦ "… good artists borrow; great artists steal." But that's another agenda. Taking is different than…

● Can you steal an idea? Are ideas something…

◈ That's a good question. That's a really good question.

✦ I don't think you can, because ideas are out there. They belong to everybody.[20]

20 See introduction.

● When people are stealing now, they are stealing an idea. They're not stealing the thing itself. They're not running over to Jeff Koons's studio and ripping the thing off the wall.

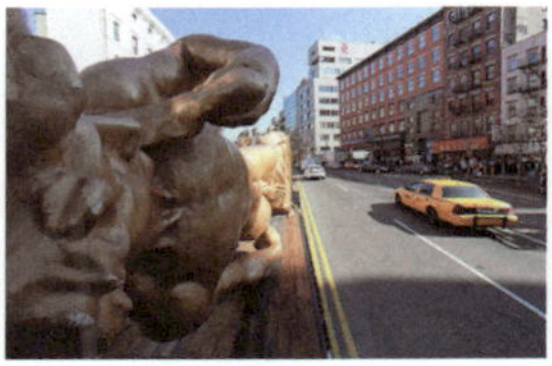

⬢ Ideas are not just floating there; someone had certain references, certain moments in their lives, certain skills, and then they built something that could be the draft of an idea – it could be a building or a sculpture. It's not that they are hanging their ideas, so that everyone can touch them, but I definitely think you can steal an idea and then produce it so that you are the author.

● No. Okay, but in writing it's different. Well, maybe it's not different, but let's say I hear that somebody has an idea. And I like the idea and then I write it. The way I would write it would be different than…

⬢ Sure.

- ○ I'm not that other writer. I couldn't write it word for word, or rather, I wouldn't write word for word the story as someone else might write it.
- ● But here again comes the issue of replication. You would develop this person's idea differently from how this person would have developed it, but then, should he choose to write it as well, he will have copied you. So, instead of being the original he will be the copy.
- ○ Who did it first?
- ● You did it first. But the idea was his.
- ✦ But who utilized it? I would like to look at it in a more pragmatic way. Everybody has an idea and ideas are free to share. I don't think you can really own an idea. Working for it, like writing a whole book and then putting it out there is taking responsibility, and therefore, authorship.
- ◈ I'm thinking of an analogy – and this might be a stretch – with Kurosawa's *Rashomon*.[21] The rape is the idea. And everybody is viewing this rape; they all perceive this. And they come back, and they tell the story, and it's ever so different than everybody else's story. And it is that filter through which we see things. Whatever idea is out there is for our taking. Whatever we do toward understanding that idea, digesting it, and putting it out again changes it and makes it ours.
- ○ Do you think it's different in other fields? Do you remember in the case of HIV discovery there was a guy in France who also claimed

21
In *Rashomon* (1950), Akira Kurosawa worked in close collaboration with cinematographer Kazuo Miyagawa. The movie tells the bandit's story, the wife's story, the samurai's story, and the woodcutter's story, which are drastically different viewings of the same incident.

22
Anthony S. Fauci (b. 1940). An immunologist who has made substantial contributions to research in the areas of AIDS and other immunodeficiencies. In 2007, he shared the Lasker Award with Carpentier and Steinman for their work in fighting HIV.

23
In 1998, author and poet Barbara Chase-Riboud accused Steven Spielberg and his production company of "brazenly" stealing from her work to make the movie *Amistad*. Later on, she dropped her plagiarism suit against the company and praised the movie as "a splendid piece of work."

to have made the discovery?[22] There was a dispute and finally, I think, they shared the credit. But in writing, there is always an end. There is a beginning and an end, technically speaking, regardless of whether one wants to operate against narrative or whatever. There are just so many stories; it's the execution that's different.

● It's really interesting what you said before about HIV because then it becomes a thing where you can claim the copyright. If you are the first person to do something, or the first person who thinks about copyright, then this initial discovery means everything – this is particularly valuable in pharmaceutical discoveries.

○ With certain discoveries you can claim copyright, yet can't patent.

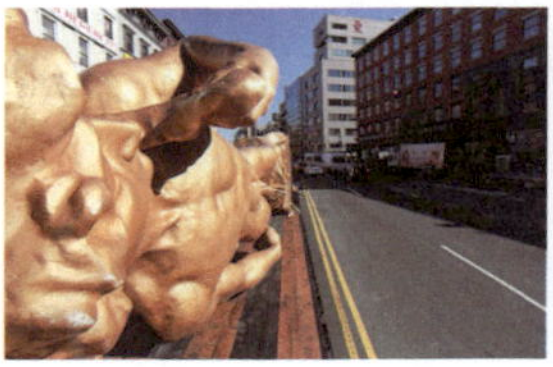

✦ Also stories for scenarios, like film scenarios…

● You copyright them?

✦ You have to.

○ Let's say that the screenwriter is in the *Guild of America*; on the West Coast, for example, there are certain regulations…

◈ Somebody had questioned Steven Spielberg about his originality in a particular case and suggested that someone else had that story.[23] And they found that there are certain stories that are just part of the public domain and over which no one can claim ownership.

✦ I asked a lawyer, and he told me that *David* is public domain. You can quote, replicate, duplicate, etc.

○ Well, certainly that happens in…

● … literature.

○ Copyright is lost fifty years after the lifetime.

● Only?

○ I think it is now seventy years – it's lifetime of the author plus seventy years when something goes out of copyright.

● And then what happens?

○ Anyone who wants to publish it can. The thing I don't like is that people might shift it. You take a novel and start putting other things in it. And I don't care if they imitate *Jane Eyre*, or something, but I don't want them to pass it off as if it wasn't written by Brontë.

● Really, that happens?

○ Well it could. Probably it doesn't happen…

● If you take *Quixote* and Cervantes it's not…

○ But if you pass it off as written by Cervantes and you have Don Quixote in the end wearing Prada, you know what I mean? What are you thinking?

✦ Don Quixote.

○ I mean I make reference to him all the time. What about translation? Everything is translation.

✦ I keep thinking about music. I find music in the middle of this discourse. In literature what you write is the book – it's not the manuscript; it's the printed version. There is no discrepancy between one copy and another. But in the music, where does the music exist? Where does the translation occur? The composer writes the symphony. Today, perhaps it's much simpler with, for example, a DJ's composing techniques. But still, there is the written music, so there is notation, and then there is the music itself – the sound in the space.

○ Yet, it's improvisation.

24
Goldberg Variations by Bach was the 1955 debut album of the Canadian pianist Glenn Gould. This album established the *Goldberg Variations* within the contemporary classical repertoire. Shortly before his death in 1982, Gould rerecorded the *Goldberg Variations*. Here, he largely abandoned the showmanship of the 1955 performance and replaced it with a more introspective interpretation that included more calculated phrasing and ornamentation.

25
Refers to *Genius Within: The Inner Life of Glenn Gould* (2010), a documentary by Canadian filmmakers Peter Raymont and Michèle Hozer.

26
Refers to Canadian sound poet, Christian Bök, who has performed an extremely condensed version of the *Ursonate* by Kurt Schwitters.

27
Wergo has published Schwitters's version of *Ursonate* in 1995.

◈ There is the pianist Stanislaw Richter, who changes up the music into his own.

✦ Also Glenn Gould. You could say his has a very strong interpretation, but beneath that interpretation, there remains a certain composed piece we agree upon.

⬢ But with literature it would be the reading, no?

✦ With literature it's…

◈ Sorry to interrupt, but if you listen to the *Goldberg Variations* by this Canadian guy, Gould… he played them when he was eighteen, then he played them six months before he died. In terms of interpretation…[24]

◉ Do they sound different?

◈ Oh yes, firmly. The later version is so soft and sensitive. And the one when he is a young man is filled with testosterone. It's just like passion.

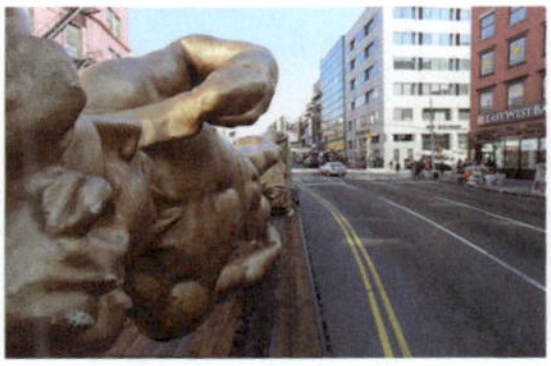

⬢ Have you seen the documentary? There was a documentary.[25]

✦ My friend Sina from *Cabinet* magazine told me a crazy story about a guy[26] who had his own issue with interpretation. Do you know Kurt Schwitters's *Ursonate*? *Ursonate* is like the first language; music of the first language. A contemporary poet tried to interpret Kurt Schwitters's text. And he was unaware of Schwitters's own recordings of the piece, or they were lost or were not found…

◉ Lost or not found?

✦ I have a CD of Kurt Schwitters reading *Ursonate*[27] with all the "*dll rrrrrr beeeee bö*" sounds, or something to that effect. It is the first language that he tries to voice. But then, before this recording was out, this other artist had the poem and he did a version of it. And then, once he listened to the original recording, he realized his version was twice as fast. It sounded like:

"dlrbebö". It wasn't human. It was so fast and hard to do. And he had no idea that Kurt Schwitters's version was much slower than his.

- ○ There was something I was just thinking about; before you started about music.
- ● Translations?
- ○ No, with writing. You could argue that the original in writing, with the book – which is a copy of a manuscript but it doesn't really matter because a book is a book – is like the response theory; it only exists when it's read. And in some sense, I really do believe that, because with a book it's this relationship between the reader and the text, and the reader is doing the interpretation, which always changes. Writing books is so different from all the other art forms because of that.
- ● Why?
- ○ Because it's always known that it's about itself. And the art world is just coming into that in the last forty or fifty years.
- ✦ With music, it's the same. The first stereo recording was made in the 1950s. The first form of replication in music was the player piano at the turn of the century. Before that, it was just the notation and the musicians playing the instrument – kind of like reading, or rich people having an orchestra.
- ○ No, but reading is different because each reader is different. Each reading is original.

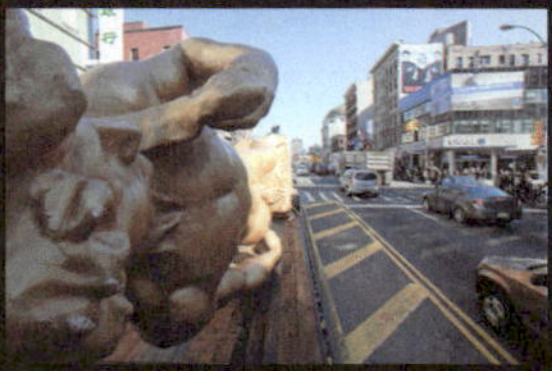

IMAGE CREDITS

FRONT COVER & BACK COVER

Serkan Özkaya, *David (inspired by Michelangelo).* Statue under construction in government building (also under construction). Istanbul, Turkey, 2005. Photo by Baris Ozcetin (Wikimedia Commons).

FLIPBOOK IMAGES

pp. 1–163. Images of Serkan Özkaya's *David* being transported through New York City, 2012. Photos by Brett Beyer.

MANIFESTO FOR THE RIGHTS OF ARCHITECTURE

p. 10
1. Adolf Loos. Josephine Baker House. Original Model, cardboard. Paris, 1927. Photo courtesy of Albertina, Vienna.

p. 10
2. Hallstatt, Austria. Photo by Philip Bouchard (Flickr Creative Commons).

p. 11
3. Ines Weizman and Andreas Thiele. Reenactment of Josephine Baker House. Ordos, Mongolia, 2008.

p. 11
4. Developer rendering for copy of the Austrian village of Hallstatt in Guangdong, China.

p. 12
5-1. Le Corbusier. Villa Savoye. Paris-Poissy, 1929. Photo by Rory Hyde (Flickr Creative Commons).

p. 12
5-2. ARM. Institute of Aboriginal and Torres Strait Islander Studies, Canberra, Australia, 2001.

p. 12
6-1. Ludwig Mies van der Rohe. German Pavilion, International Exhibition. Barcelona, 1929.

p. 12
6-2. Reconstruction of Ludwig Mies van de Rohe's German Pavilion. Barcelona, 1986. Photo by Le.Mat (Flickr Creative Commons).

p. 12
7-1. Foster + Partners. Masdar City. Abu Dhabi, UAE, 2007. Image courtesy of Foster + Partners.

p. 12
7-2. OMA. RAK Gateway. UAE, 2007.

p. 12
8-1. Le Corbusier. Pavilion de L'Esprit Nouveau in Exposition Internationale des Arts Decoratifs et Industriels Modernes. Paris, 1925.

p. 12
8-2. Reconstruction of Le Corbusier's Pavilion de L'Esprit Nouveau. Bologna, 1977. Photo by Enrico Celesti (Flickr Creative Commons).

p. 13
9-1. St. Peter's Basilica. Rome, 1626.

p. 13
9-2. Basilica of Our Lady of Peace of Yamoussoukro. Ivory Coast, 1985–90.

p. 13
10-1. Doge's Palace. Venice, 1309–24. Photo by Clayton Parker (Flickr Creative Commons).

p. 13
10-2. Doge's Palace in the Venetian Macau Resort Hotel. Macau, 2002. Photo by Ken Wilson Lee (Flickr Creative Commons).

p. 13
11-1. Frauenkirche. Dresden, 1760.

p. 13
11-2. Reconstructed Frauenkirche. Dresden, 2005. Photo by Sven Wusch (Flickr Creative Commons).

p. 13
12-1. Stari Most. Mostar, Bosnia and Herzegovina, 16th century. Photo by Hannes Grobe (Wikimedia Commons).

p. 13
12-2. Rebuilt Stari Most. Mostar, Bosnia and Herzegovina, 2004. Photo by Duncan Hill (Flickr Creative Commons).

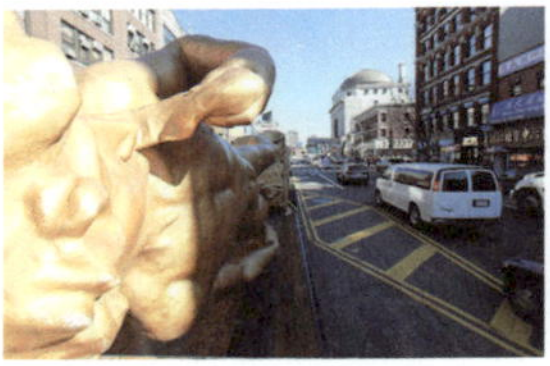

HOLY HOLY COPY COPY CULTURE CULTURE

p. 19
1. Anonymous (School of Leonardo da Vinci). *La Gioconda o Mona Lisa.* 1503–16. Image courtesy of Museo Nacional del Prado. The painting is the earliest known copy of the *Mona Lisa* by Leonardo.

p. 20
2. Leonardo da Vinci. *Mona Lisa.* 1503–17. Photo of painting in the Louvre by Mark Squires.

p. 21
3. Postcards of *Mona Lisa.* Photo by John Perivolaris (Flickr Creative Commons).

IMAGE CREDITS

p. 22
4. Stills from *P. Diddy vs. Mona Lisa*. Incubate blog. 2010. http://inlog.org/2010/10/18/jay-z-vs-mona-lisa/

p. 23
5. Copy of *Mona Lisa* in Dafen, China. 2011. Photo by Michael Mandiberg (Flickr Creative Commons).

p. 24
6. Ursito. *Mona in Red Dress*. 2006. Image courtesy of Mega Mona Lisa.

p. 25
7. David Shankbone. *Friday, Day 14 of Occupy Wall Street*. 2011. (Wikimedia Commons).

p. 26
8. Nevit Dilmen, *Symbol of Kopimism or Kopimizm, including Ctrl-C, Ctrl-V, Yin yang*. 2012. (Wikimedia Commons).

ANIMATED DOUBLES

p. 34
1. Serkan Özkaya. *A Sudden Gust of Wind*. Bilsar Building, Istanbul, 2009. Photo by Baris Ozcetin.

pp. 35, 36
2. Jeff Wall. *A Sudden Gust of Wind (after Hokusai)*. Transparency in light box. Tate Gallery, London, 1993. Image courtesy of Tate Gallery.

pp. 37, 38
3. Katsushika Hokusai. *Ejiri in Suruga Province* (A Sudden Gust of Wind). Color woodcut. 1830–33. Image courtesy of Trustees of the British Museum.

p. 39
4. Serkan Özkaya. *A Sudden Gust of Wind*. Boots Contemporary Arts Space, St. Louis, Missouri, 2008. Image courtesy of Boots Contemporary Arts Space.

p. 45
5. Detail from Sandro Botticelli's *The Birth of Venus*. Tempera on canvas. Galleria degli Uffizi, Florence, c. 1482. Image courtesy of Galleria degli Uffizi.

p. 46
6. Serkan Özkaya using a spray can of instantly drying adhesive agent (glue). Photo by Baris Ozcetin.

p. 47
7. Detail (wind gods) from Botticelli's *The Birth of Venus*.

p. 56
8. Serkan Özkaya, *David (inspired by Michelangelo)*, (aborted). Installation in Sishane Square, Istanbul, Turkey, September 2005. Film Still from Danila Cahen. *Friendly Enemies*. 2009.

pp. 57, 58
9. Serkan Özkaya, *David (inspired by Michelangelo)*. Statue under construction in government building (also under construction), Istanbul, Turkey, 2005. Photo by Baris Ozcetin (Wikipedia Commons).

pp. 59, 60
10. Film still of a model of David against the Hagia Sophia, Istanbul, from *Double David*. Dir. Gabriel Wrye. Zoon Politicon. 2012.

pp. 61, 62
11. Serkan Özkaya, *David (inspired by Michelangelo)*, on top of a motor truck, drawing. 2012.

p. 63, 64
12. Serkan Özkaya, *David (inspired by Michelangelo)*, in front of Lincoln Center, NYC, March 2012. Photo by Brett Beyer.

p. 65
13. Serkan Özkaya, *David (inspired by Michelangelo)*, installation in Louisville, Kentucky, April 2012. Courtesy of 21c Museum. Photo by Magnus Lindqvist.

pp. 68, 69
14. Film stills from *The Student of Prague*. Dir. Paul Wegener. 1913.

DOUBLE (A THEORY)

p. 72
1. Allan McCollum. *Plaster Surrogates*. Enamel on Hydrostone, sizes variable, each unique. 1982–84. Installation: Metro Pictures Gallery, New York, 1986. Photo courtesy of Allan McCollum.

p. 73
2. Film still from *A Film About a Woman Who*. Dir. Yvonne Rainer. 1974.

p. 74
3. Slater Bradley. *I Hate Myself and Want to Die*. C-print. 2004. Image courtesy of Slater Bradley.

p. 76
4. Slater Bradley. *Doppelganger in Times Square*. C-print with crushed moon gold leaf. 1999–2010. Image courtesy of Slater Bradley.

IMAGE CREDITS

p. 77
5. Conrad Bakker. *Untitled Project: Used Books [Assorted].* Oil on carved wood. 2008. Image courtesy of Conrad Bakker.

p. 78
6. Conrad Bakker. *Untitled Project: SIGN [Relax and Take Your Fucking Time].* 2010. Image courtesy of Conrad Bakker.

MEMOS: ON DOUBLES, TRIPLES, AND SERIALITY

pp. 82, 83
Yen-Wei Liu. *Collage: Repeat and Serial Offenders (Arbus, Glass, Khafre, Khufu, Judd, Leen, Menkaure, Stella).* 2013.

A CALL FOR THE REENACTMENT OF SOMETHING THAT NEVER ACTUALLY HAPPENED

p. 90
Thumbnail I. E.T.A. Hoffmann. Instantaneous ink drawing for his book, *Sandman.* 1815. (Wikimedia Commons).

p. 90
Thumbnail II. Damien Hirst. Dot painting for Beagle2 Calibration Target Plate. 2003.

p. 92
Thumbnail III. Paul Cocksedge. *A Gust of Wind.* Installation with Corian plates at the London Design Festival. 2010.

p. 92
Thumbnail IV. Detail from Paul Cocksedge's *A Gust of Wind.*

p. 94
Thumbnail V. Fake Industries Architectural Agonism. The reconstructed walls of Ai Weiwei's studio. 2011.

IN THE FUTURE THE DOME SHOULD BE A PLACE WHERE PEOPLE CAN FUCK FREELY

p. 98
Thumbnail I. Eduardo Souto de Moura. House in Sierra da Arrabida. Portugal, 1994–2002. Photo by Luís Ferreira Alves.

p. 100
1. Jackson Pollock. *Number 8.* Oil on canvas. 1949.

p. 100
2. Francis (Frankie) Brown with one of his "splatter-dash" oil paintings. 2006.

p. 102
Thumbnail II. Cover of Kenneth Goldsmith's *Soliloquy.* 2001.

p. 103
3. Coverage of Özkaya's *David (inspired by Michelangelo)* in the New York Times. 7 March 2012.

p. 105
Thumbnail IV. Title page of the 4th edition of Miguel de Cervantes' *Don Quixote.* 1605.

p. 105
Thumbnail V. Cover of the first edition of James Joyce's *Ulysses.* 1922.

p. 113
4. Storefront for Art and Architecture. *Past Futures, Present, Futures.* 2012.

p. 115
5. Buckminster Fuller. *Dome Over Manhattan.* 1960. Image Courtesy of the Estate of R. Buckminster Fuller.

LET'S ASK THE QUESTION DIFFERENTLY: CAN YOU HAVE COPIES IF THERE'S NO ORIGINAL?

p. 128
Thumbnail I. Tullio Lombardo. *Adam.* Marble. 1490–95. Image courtesy of The Metropolitan Museum of Art, New York.

p. 136
1. Gillian Steinhardt. David's penis and fig leaf necklaces. Brass. 2012.

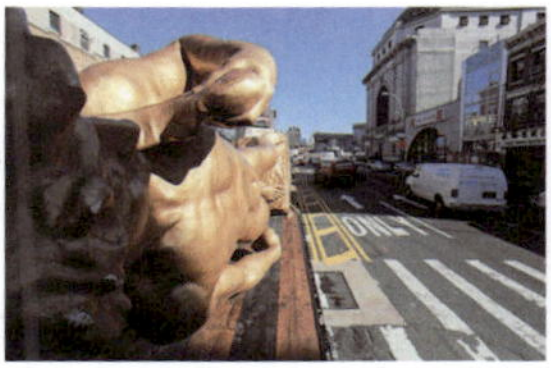

p. 145
2. Louis Daguerre. *Boulevard du Temple, Paris, 3rd arrondissement.* Daguerreotype. April 24, 1838 to May 4,1838. (Wikimedia Commons). This image is purportedly the first picture of a living person.

p. 147
3. Ludwig Mies van der Rohe. German Pavilion, International Exhibition, Barcelona. Inaugurated in 1929, demolished 1930. Photo by grundkonzept (Flickr Creative Commons).

p. 149
Thumbnail II. Han van Meegeren. *Lady Reading Music*. Oil on canvas. 1935–40. Rijksmuseum, Amsterdam.

p. 149
Thumbnail III. Johannes Vermeer. *Woman in Blue Reading a Letter*. Oil on Canvas. 1663–64. Rijksmuseum, Amsterdam.

p. 150
Thumbnail IV. Computer generated 3D print-out of Michelangelo's *David* by Gentle Giant Studios in CA. After a computer model by Dr. Marc Levoy. 2004.